AN OUTLINE OF BRITISH CROP HUSBANDRY

AN OUTLINE OF BRITISH CROP HUSBANDRY

BY

H. G. SANDERS, M.A., Ph.D.

Chief Scientific Adviser (Agriculture)
Ministry of Agriculture, Fisheries and Food

CAMBRIDGE
AT THE UNIVERSITY PRESS
1958

CAMBRIDGE
UNIVERSITY PRESS

University Printing House, Cambridge CB2 8BS, United Kingdom

Cambridge University Press is part of the University of Cambridge.

It furthers the University's mission by disseminating knowledge in the pursuit of education, learning and research at the highest international levels of excellence.

www.cambridge.org
Information on this title: www.cambridge.org/9781107586574

First edition 1939
Reprinted by offset litho 1943
Reprinted by offset litho 1945
Second edition, reset & revised 1949
Third edition, reset & revised 1958
First paperback edition 2015

A catalogue record for this publication is available from the British Library

ISBN 978-1-107-58657-4 Paperback

CONTENTS

ILLUSTRATIONS

*This map is available for download from www.cambridge.org/9781107586574

PREFACE TO THE FIRST EDITION

A complete treatise on crop husbandry, even though limited to Great Britain, would be a cumbersome volume. This book does not attempt to present the whole picture, but only to sketch in the outline: therefore, important subjects such as varieties, yields and uses of the common crops have been omitted. In particular, it may be urged as a serious criticism that grass—the most important crop in Britain—is only referred to incidentally. But with that crop there are many important details of husbandry which are peculiar to itself, and hence are best dealt with in books devoted entirely to grassland. To a greater or lesser extent this is true of all crops, and consequently the majority of published works have been concerned with the different crops severally; it was with the object of providing some sort of link between these works that the present book was written.

Farming conditions vary widely even within the confines of this small country, and a writer inevitably gives undue weight to the systems and methods common in the district with which he is most familiar. It must be admitted frankly that crop husbandry is viewed in this book through East Anglian spectacles, but it is hoped that great distortion will not be found, because the aim has been to subordinate detail to principle: the former varies sharply from district to district, almost, indeed, from parish to parish, but the latter should have general applicability. Undoubtedly, success in farming demands a thorough mastery of local detail, but that can soon be added to a knowledge of the underlying principles, and of the manner in which they are applied elsewhere.

It gives me great pleasure to acknowledge my indebtedness to Mr A. Amos, M.A.; his generous assistance, when I took over from him a course of lectures on crops, laid a deep obligation upon me. To Mr W. S. Mansfield, M.A., and to Mr F. H. Garner, M.A., I am particularly grateful, not only for all that I have learnt from them during ten years of very pleasant association in the teaching of agriculture, but also for their reading of the whole of the typescript; their helpful criticisms and suggestions have been invaluable and greatly appreciated. My grateful thanks are also due to Mr F. A. Buttress, who has aided me in selecting and checking the references given at the end of chapters, and to Mr D. A. East for his almost uncanny skill in reading my handwriting, and for checking the Index.

H. G. SANDERS

June 1939

NOTE ON THE SECOND EDITION

Although the whole book has been revised for the Second Edition, it is only in Chapter XII that major change has been made. Mature reflection and kindly instruction from Professor Edgar Thomas have convinced me that theoretical estimates of costs of production are not worth the paper on which they are written. This chapter has, therefore, been shortened very considerably.

H. G. S.

March 1949

NOTE ON THE THIRD EDITION

A new edition has given the opportunity to bring the book up to date, to incorporate the very large advances that have been made in crop husbandry during the last ten years. New materials and new methods have become available to the farmer, who has had to find how best to utilise them in his scientific art, but nothing has occurred to change fundamental principles. It has not, therefore, been necessary to alter the general layout of the book. The difficulty has been to decide what new has become established, and so should be described in a textbook, and what is promising but still unproven.

One change that cannot be ignored is the continuing increase in the use of compound fertilisers. Somewhat reluctantly it was decided to discuss manurial dressings in terms of plant nutrients rather than of straight fertilisers. In past editions a short guide to further reading was given at the end of each chapter but this has been discontinued. Publications of merit follow each other very closely so that numbers of references would be very great, for few of the older ones could be dropped; this, of course, is the ever more harassing problem of the teacher.

I am greatly indebted to Miss L. Samiloff for very considerable help in the onerous work of typing and checking.

H. G. S.

August 1957

CHAPTER I

ROTATIONS

A study of crop husbandry logically starts with a consideration of rotations, because the treatment that a field has received in previous years greatly affects the success of a crop which, in its turn, may have an important influence on production in succeeding years. No true picture of the right methods of husbandry can therefore be gained if each crop is regarded as a separate entity: from the outset the view to be cultivated is the long one, which embraces the effect of one crop upon another, and the considerations which should guide a farmer in determining his sequence of cropping.

At the dawn of history, when man forsook the chase for the plough as his main means of obtaining food, his principle was to crop a convenient piece of land continuously until it was exhausted, and then to abandon it and move to a fresh site. This was undoubtedly sound as long as fresh sites were available, but is scarcely to be recommended in the modern world. Nevertheless, it is still the general procedure of natives in certain parts of Africa, and the continuous growing of wheat on American prairies was essentially similar. During the agricultural depression between the two World Wars some of the poorest lands of this country were abandoned to whatever nature ordained should grow there. In the food production campaign of 1939–45 these derelict areas were reclaimed and, with proper draining, liming and manuring, produced very satisfactory and welcome crops. Whilst they were rough they slowly accumulated a small store of fertility, exemplifying the underlying principle of shifting cultivation.

Any permanent form of agriculture clearly requires a sequence of cropping which will maintain the land continuously in a cultivable condition. Such a rotation was first established in this country in the early centuries of the Christian era, and it persisted over a large part of England with very little variation for over 1500 years. The rotation—autumn corn, spring corn, fallow—was almost universal for that long period, during which departure from it was severely discouraged, though a few bold spirits experimented to some extent. It was not until the eighteenth century, when roots and clover became established as farm crops, that any general change was made. The introduction of these two crops enabled farmers to utilise their land more fully; on the lighter soils it was no longer necessary to waste a year in fallowing, whilst on the heavier soils the proportion of fallow could be reduced well below one-third. During the eighteenth century rotations including roots and clover were slowly evolved and gradually adopted; in general it can be said that there was little further change till the beginning of the present century.

During recent years rotations have changed rapidly, but the most fundamental change has been in the rigidity to which they are adhered. Previous to 1874 a tenant farmer had no choice in the matter. His lease prescribed the rotation on which he was to farm, and he could not depart from that sequence without his landlord's permission, which was sparingly granted. A series of Acts passed from 1874 to 1908 gave tenant farmers freedom of cropping, but they are still liable to dilapidation claims on quitting their farms if their system of cropping has impaired the fertility of their land. At the present time it seems but natural that a farmer should be allowed to grow what he likes on the land that he rents, but it is not justifiable to regard the old insistence of landlords that their tenants should keep to a stated rotation as merely tyrannical. It was not only to the landlord's advantage, but also to the tenant's, that the land should be maintained in a state of high fertility, and this was achieved very largely by a good rotation. It must be remembered that chemical manures have only become available quite recently, and knowledge of their potentialities is still not universal among the farming community; now that sufficient supplies are available an informed farmer may, without impoverishing his soil, take liberties which would have been disastrous for his forefathers. Again, the tractor has given the farmer power to work the soil more thoroughly at crucial seasons, and with the further aid of modern selective weed killers has made it possible to keep the land reasonably clean of weeds, even though corn be grown more frequently than an earlier generation of farmers would approve. Furthermore, an immutable rotation is only suitable for settled economic conditions, and since 1914 conditions have been anything but settled. It is not surprising, therefore, that rigid adherence to rotations is no longer common, that progressive men are often heard to speak with scorn of rotations, and that the present generation is seeing bold experiments made in this connection.

It would be idle to deny that many men, who have cropped their land with an eye on market probabilities rather than in conformity with established practice, have been successful; at the same time it is undoubtedly true that some who have boasted that they were untrammelled by tradition have been forced to revert to the old-established rotations. It is best to keep a very open mind on this question of originality versus custom. Conditions have changed so much in the last hundred years that old sequences of cropping may be out-of-date, but few farmers will be in a hurry to discard that which is sanctified by centuries of experience; in this they are wise, for it must be remembered that weaknesses in rotations show themselves very slowly. Thus, despite the fact that good farmers often boast that they are no slavish followers of a rotation, it is as important as ever to have a clear understanding of the principles underlying rotations; perhaps it is even more important, since if one leaves the beaten track it is more necessary to know the location of pitfalls than if one keeps to the broad highway.

It will have been gathered from what has been said above that a rotation is a settled sequence of cropping; as will be seen later, its length may be almost anything, but in general it is between 3 and 8 years. To illustrate how a rotation works in practice, a moment's consideration may be devoted to a very well-known case—the Norfolk four-course rotation of roots-barley-seeds-wheat. Approximately one-quarter of the arable land will grow these crops in the 1st, 2nd, 3rd and 4th years respectively, returning to roots in the 5th, barley in the 6th, and so on. Another quarter will follow exactly the same course, but will start one year ahead, growing barley in the 1st year; a third quarter will carry seeds in the 1st year, whilst the remaining quarter will carry wheat. In this way the acreage of the various crops on the farm remains roughly constant from year to year, but each field grows all crops in succession.

UNDERLYING PRINCIPLES

The principles on which rotations have been built up are very diverse; some of the principles are based on science, some on economics, whilst others are matters of common sense. It will be convenient to divide their consideration into two parts, and to deal first with the advantages gained from growing crops in a rotation, and then to take the points which arise in deciding on a rotation for a particular farm.

ADVANTAGES OF A ROTATION

1. The fact that the farmer is working to a settled plan, and is not constantly improvising. In view of the varied activities and interests of the farmer, and of the way in which he has to dovetail his work and that of his men, the point is entitled to some weight. It may be wholly admirable for engineers to improvise, because they thoroughly understand the construction and working of the machines under their control, but the farmer is dealing with biological material, and his knowledge of its working is, by comparison, miserably superficial. If he keeps to a rotation carefully thought out in the light of other people's experience he has little to fear, but if he strikes out boldly for himself it may be equivalent to putting a spanner among the cog wheels. Furthermore, a settled plan of cropping may be accompanied by a more or less settled plan of manuring; advice on the plan, if necessary, can easily be obtained, and adherence to it (with, of course, variations dictated by level of fertility of particular fields, by prices, etc.) will maintain at a roughly constant level the fertility of all the arable land.

2. All the other advantages of a rotation follow from the fact that it ensures that each field receives a proper succession of crops. The points to be considered here are of paramount importance, and it will be easier to take them under three headings.

(*a*) *Cleaning.* There are many species of weeds, and between them they exhibit almost every conceivable habit of growth. It follows that if the same, or a similar, crop be grown continuously on the same piece of land there will be at least one species of weed that will be favoured by the time and type of cultivation carried out; this species will therefore spread rapidly, and from being perhaps relatively innocuous will become so prevalent that it will dominate the crop. As an example, the common infestation of heavy land with slender foxtail (commonly called black grass) may be mentioned. This weed is an annual, germinating from October to Christmas, and it ripens its seeds in June and July. On heavy land autumn sowing is very common, because farmers know from experience that on such land it may be impossible to drill in spring; thus in a large proportion of years a seedbed is prepared at the optimum time for the slender foxtail to germinate, and a corn crop is planted which will not be harvested until after the weed has ripened and shed its seed. To control the weed a very good method is to go through the motions of preparing an autumn seedbed, but not to drill the seed; then, when the slender foxtail has germinated, the land may be cultivated, or even ploughed, and the young seedlings killed in vast numbers. In other words the method rests in risking a wet spring and taking a spring-sown crop, this being more effective than late sowing of an autumn crop. This is only one example, but many others might be quoted to show that a rotation, by bringing different crops to a field in successive years, and thus ensuring cultivation at different times of the year, does much to prevent the land from becoming foul. Continuous wheat growing has often been attempted but has always broken down because the land has become foul.

There is another way in which a rotation helps to keep the farm clean. It will ensure that fallows, cleaning crops, semi-cleaning crops and smothering crops occur in due course on all the fields of the farm; the fact that this point is obvious should not blind the reader to its supreme importance.

(*b*) *Diseases and Pests.* Many and varied are the diseases and pests to which crops are subject, but most of the ills are specific to one or two crops; where, therefore, the organism lives in the soil, the first method of control is not to grow susceptible crops frequently on the same field, and thus to avoid providing the organism with hosts on which to multiply. It is true that some farmers have neglected the danger with impunity, but they do so at great peril, and many grim tales can be told to show how real the danger is. Eelworm provides an example. Eelworm may be present in a field in considerable numbers without any appreciable harm to a crop, but if the numbers of a particular strain of eelworm exceed a certain figure (which may be of the order of 200,000,000,000 per acre) practically complete failure of crops susceptible to that strain results. It is a merciful fact that the different strains of eelworm rigidly adhere to one or two crops each as their victims, and consequently ordinary good farming—of which growing crops in a rotation forms a part—is sufficient to keep the numbers down; but there have been cases, where rotations have been discarded, where the most serious results have occurred. There is an area of greensand in Bedfordshire eminently suited by its soil and situation to market gardening and, in particular, to the production of early potatoes, and that crop has been taken continuously on the same field by a number of farmers; the potato eelworm has increased on some of these fields so much that it has become impossible to grow a crop of potatoes at all. The land is practically pure sand and must be farmed very highly if a crop is to be grown, and only early potatoes appear to justify the generous treatment necessary; when that crop becomes impossible the land is not worth farming and consequently fields, which previously commanded rents of £3 per acre or more, have been allowed to go derelict. Then the eelworm slowly dies out and after 12 or 15 years the land is no longer 'potato sick' so that it can be brought back to production again; there arises again the temptation to crop too closely with potatoes and thus to build up once more an overwhelming population of eelworm. The fact that the severe infestation with eelworm was simply due to bad farming has been recognised in law, and heavy damages have been awarded against a farmer for ruining his landlord's property by growing potatoes too frequently. It is not only in Bedfordshire that the trouble has arisen; there are many other districts in the country which could be quoted to show how neglect of the principle of rotations, by growing potatoes too frequently, has led to disaster.

The sugar-beet eelworm has also had devastating effects, and some sugar-beet factories have had to close down in Germany because the area they served became so badly infested with the pest, through too frequent growing of the crop. It is a reproach to British farming that within 15 years of the introduction of sugar beet to general cultivation crop failures due to eelworm occurred. The peril was appreciated by the sugar-beet factories and a clause was introduced into the farmers' contract with the factory, prohibiting the growing of sugar beet for 2 years running on the same field; this was a wise clause but not sufficiently stringent and in 1943 the Sugar Beet Eelworm Order was made to give a tighter control. The latest Order (1952) prohibits, save under special licence from the County Agricultural Committee, the growing of sugar beet, mangolds, red beet, spinach and the brassicae on any land where sugar-beet eelworm is known to exist, and on all land within an Infected Area where any of those crops have been grown during the two preceding years. In the schedule attached to the Order nine Infected Areas are listed—a large one embracing parts of the counties of Cambridgeshire, Huntingdonshire, Isle of Ely, Norfolk and West Suffolk: a second including parts of the Soke of Peterborough and of the Holland Division of Lincolnshire: and seven sewage farms, those of Loughborough, Norwich, Haverhill, Bury St Edmunds, Higham Ferrers, Rushden and Wisbech; others may be added from time to time. The brassicae are included because their growth on infected land builds up the population of sugar-beet eelworm, although they do not themselves succumb to the pest. A further stipulation of the Order is worth noting, as it shows how easily the eelworm can be spread. Potatoes grown on land in which sugar-beet eelworm is known to exist may not be sold for planting on any other farm; the danger is that the sugar-beet eelworm may be in the soil adhering to the potato tubers. There are other strains of eelworm which attack different crops, but the point need not be laboured further. Eelworm is not the most common pest in the country and is mentioned here only as an example of the danger of growing the same crop too frequently on the same field; it must be clearly stated that there are numerous other insect pests, and many fungi, which may cause serious trouble if the principles of rotation are flouted. It is important to realise that similar crops may be susceptible to the same dangers, so that a change from one to the other may not tend to check a disease; for instance, wheat and barley are both susceptible to take-all.

(*c*) *Maintenance of Condition.* As applied to land the word 'condition' is frequently, but improperly, held to be synonymous with fertility. The latter is inherent and is in fact that for which the farmer pays rent. Condition is that for which the farmer alone is responsible, and is not static but cumulative. It includes the state of cleanliness of the land and its freedom from disease, but in this section it is only proposed to deal with the question of plant food. The necessity of maintaining the stores of plant food in the land by a reasoned sequence of cropping was, in bygone days, of the very first importance; in fact it was nothing less than a *sine qua non* of any system of permanent agriculture. Its importance has undoubtedly decreased, now that an abundant supply of nitrogen is available, but though the prices of chemical manures are reasonable, they are high enough to warrant careful economy in their use. Furthermore, too much reliance on chemical fertilisers is not to be advised, because of the danger of miscalculation, and consequent unbalance of plant nutrients in the soil.

The different crops remove from the soil varying amounts of the several nutrients; this is illustrated by Table I, which gives the amounts of the four chief plant nutrients removed from the soil per acre, by six different crops. The yield and also the composition of a crop vary so much that the figures can only be accepted as representing very approximately the drain on the soil, but they do suffice to show the fact that the plant food removed varies widely from crop to crop. There is clearly a danger that the too frequent growth of one crop may so denude the soil of the available part of one plant nutrient that it becomes the limiting factor in the condition of the field.

The different crops have root systems of varying depth, and consequently a rotation, by bringing various crops to a field in turn, ensures, as far as is possible, that the upper and lower layers of soil shall all be searched by roots for the plant nutrients they contain. Then it must be remembered that some crops respond generously to, and therefore justify, heavy manuring, and a rotation ensures that the heavily manured crops shall not tend to occur too frequently on a few fields, but shall be equally spread over the farm, and so maintain the general level of condition. Finally, some crops require more thorough and deeper cultivation than others; it is believed that good cultivation is beneficial, by promoting aeration and thus making unavailable plant food available, so that in this way, too, a rotation may help to maintain the general condition of the farm.

TABLE I. PLANT NUTRIENTS REMOVED BY DIFFERENT CROPS

Crop	Amount removed per acre (lb.)			
	Nitrogen (N_2)	Phosphate (P_2O_5)	Potash (K_2O)	Lime (CaO)
Wheat: 40 bushels and 30 cwt. straw	60	26	42	11
Beans: 28 bushels and 20 cwt. straw	88*	23	66	30
Red clover: 2 tons hay	97*	17	99	72
Swedes: 15 tons roots	71	27	101	27
Sugar beet: 10 tons roots and 7 tons tops	90	35	169	80
Potatoes: 8 tons tubers	61	32	108	5

* These, being leguminous crops, do not, of course, denude the soil of nitrogen.

POINTS TO CONSIDER IN DECIDING ON A ROTATION

(i) *Part of the Farming System*

Rotations are fundamental to crop husbandry, but crop husbandry is only a part of farming. The first question to be decided is what is to be produced from the farm, and then the crops must be selected and apportioned, as far as circumstances permit, to fit in to a well-knit system. There is little to be said for regarding a farm as a series of water-tight compartments; it may be large enough to justify some specialisation among the men, but the farmer has to look at it as a whole, and he may find it advisable to plan, or to change, his rotation with his eye more on animal, than on crop, husbandry. If, for instance, a grassland flock is kept, and it is found to be more profitable to keep the lambs for selling fat during the winter, a sufficient acreage of fodder roots must be provided to fatten them. Where an arable flock is being kept the rotation will be largely determined by the necessity of providing food for the flock in every month of the year. It is possible to feed a dairy herd entirely on purchased foods, but it is generally more profitable to produce at least the roughages in its ration on the farm. The head of livestock will determine the requirements of hay, and deficiency in the supply from the grassland must be made up by short leys on the arable land. If it is proposed to winter a number of cattle in open yards they will need a generous allowance of straw for bedding, and a mixed farmer rarely expects to buy straw. Beans and peas are the only home-grown concentrates rich in protein, and they may well find a place in the rotation as a way of keeping down cake bills. Nor is it only requirements of stock that have to be considered. In the Fens one very good reason, in some cases the chief reason, for growing wheat is to provide straw for covering potato clamps. The widest possible view of the whole undertaking must be the starting point, and this will usually show that certain minimum requirements of particular crops should be met, and so provide important guidance in determining the rotation.

(ii) *Suitable Crops*

The success of a farming venture clearly demands that the crops sown should be suitable to the situation of the farm, in regard to soil and climate. The suitability of a crop to a particular soil type is by no means a clear-cut matter. Of most crops there are a number of varieties in this country, and the varieties often show predilections for different soils. Sugar beet grows well on heavy land, and in a dry summer may do better there than on light soil; but the difficulty of getting the roots out of the ground, the labour of carting the crop off in the autumn, and the resulting poaching of the soil, make it unsuited to clay. Roots for winter folding are restricted to light land, chiefly because it is only on such soil that it is possible to fold them during winter.

A list showing the suitability of crops to different soils must not, therefore, be regarded as definite and clear-cut; with this proviso the following may be given for the common crops:

Lightest sands: rye, lupins, carrots, kidney vetch.

Light soil: barley, roots for folding, sugar beet, potatoes, peas, market-garden crops, short leys.

Light chalks: as above except for potatoes. Sainfoin, trefoil and lucerne very suitable.

Medium loam: practically all crops.

Heavy clay: wheat, oats, beans, mangolds, leys and permanent grass.

Fens and silts: wheat, potatoes, sugar beet, market-garden crops (especially celery), root crops for seed, buckwheat.

Acid soil: oats, rye, potatoes.

In regard to the last, it must not be thought that oats, rye and potatoes grow best on acid soil. They will give full crops where there is a slight degree of acidity but they suffer markedly on very acid soil. If the pH value of a soil is low, lime should be applied at the very first opportunity, but if some crop must be grown before liming can be carried out then one of these three should be taken.

As regards climate, the crops most suitable for the wetter districts are oats, turnips and grass (long leys and permanent pasture). In the drier districts wheat is particularly suitable, whilst barley does well except in very droughty seasons; of the root crops mangolds, sugar beet and kohlrabi do best where rainfall is low, whilst leys tend to be shorter, and to consist largely of legumes which grow luxuriantly on the spring rainfall. There is a close connection between soil and climate in regard to suitability of crops. For any crop, the wetter the district the lighter the soil on which it will grow best; thus barley is generally most successful on light land, but in a very dry year it may ripen prematurely, and plumper and higher quality grain is obtained from heavy land. In any case, views on quality in barley are being modified at the present time. What has always been the aim is quality for the maltster, who requires a grain full of starch and very low in content of nitrogen. For that he has been willing to pay up to double the price that barley was worth for feed. At present the difference in price of malting and feeding barley is very small so that yield has become much more important than quality. This has widened the field for barley very considerably since high yield can be obtained on any good soil, certainly not excluding the clays.

It is not only the soil and climate which must be considered in deciding whether a crop is suitable to a particular farm; a favourable local market may be available, and this would justify the inclusion of a crop, even though the natural conditions of the farm were not appropriate. If a sugar beet factory is in the vicinity advantage will naturally be taken of the fact, and a fair acreage of sugar beet allowed in the rotation, because low transport costs will make that crop especially profitable. Another illustration of how the limitations which would normally be set by conditions of soil and climate may be overridden is provided by the large acreage of potatoes grown on heavy land in Essex, because of the immense market in the neighbourhood provided by London and its suburbs. The suitability of a crop may also depend to a large extent on the amount and skill of labour available, and this point is of great importance in market gardening and, in particular, in fruit growing.

Suitability is therefore a matter of degree. In drawing up a rotation the aim should be to include a large proportion of crops which may, on the lines indicated above, be regarded as suitable. This does not preclude a small acreage of other crops, which may be required by the general organisation of the farm or for the sake of diversity, but the success of the farm will clearly demand that the main crops are such as can be grown successfully.

(iii) *Even Distribution of Labour*

When it is remembered that a farm worker has to turn his hand to a wide variety of quite skilled jobs, and that it is impossible for the farmer or his bailiff to keep an eye on men dispersed over a farm, it will be realised how much a farmer's success depends on his men. It is very important, therefore, to collect and retain a good staff of workers. But a good man not unnaturally looks for work that is permanent, and employs him for 52 weeks in a year; he often lives in a tied cottage and he expects neither to change his employer nor to suffer periods on unemployment pay. It follows that a farmer should aim to employ a constant number of men throughout the year, except perhaps for the addition of casual labour for short periods for such work as potato planting and lifting. A good rotation should, therefore, create as constant a labour demand in all months of the year as possible; during the months in the depth of winter there will naturally be little work to do on the land, but hedging, ditching and draining, threshing and the extra care required for housed livestock, will help to find productive work at that time of the year. As an example of a system of farming which fails badly on this point, Canadian wheat farming is frequently quoted; this has been described as six months' hard labour followed by six months' idleness.

The labour requirements of four crops are compared in Figure 1, which illustrates records obtained by the Economics Section of the Department of Agriculture of Leeds University. The figures have been taken from the cost records of 1400 crops grown in Yorkshire from 1948 to 1954 and the figure illustrates the man hours and tractor hours work per acre for the calendar months of the farming year, October to September. Each crop entails some work which is done during the winter months as other work, or the weather, allows; these work units are represented to the right of the figure and it will be realised that they go far to fill the gap shown for January and February, although they do overlap partly into other months. It will be seen that winter cereals create a fairly heavy labour demand in early autumn when the seedbed is being prepared and drilled; from then till harvest these crops require very little attention. The extra work, almost entirely man hours, involved when the binder is used, as compared to the combine, is shown and following the binder there is a considerable demand for man labour for threshing during the winter period. The cultivation required for spring cereals is rather greater, and more spread out, than that for winter cereals; winter and spring cereals fit in quite well except, of course, for the busy harvest period. The man labour required per acre of a sugar beet crop is approximately six times as much, and the tractor hours are three times as much, as for cereals. The important point to notice is that the distribution of the heavy labour demand of sugar beet differs markedly from that of cereals; the intense period in autumn overlaps and partly clashes with the autumn requirements of winter cereals but the other intense period in spring and summer falls at a time when the requirements of cereals are practically nil. The total labour requirements of maincrop potatoes are not very dissimilar from those of sugar beet but, again, there are important differences in seasonal distribution. The heavy work of planting potatoes slightly precedes the spring peak for sugar beet hoeing and singling, though subsequent intertillage of potatoes clashes somewhat with that of sugar beet; potato harvesting is condensed into a short period of very intense labour demand which precedes the heaviest of the lifting work with sugar beet. The high demand for man labour for riddling potatoes is shown as winter work, but it should be realised that riddling may be spread over any or all of the months from October to May. The important point to notice from Figure 1 is that root crops create a very heavy labour demand in spring and summer, when cereal crops require very little, if any, attention; furthermore, at the busy time of harvesting cereals, root crops are established and can be left untouched. It should be very clear that when a farm carries a high proportion of one crop, or of similar crops, the labour requirements must be very uneven throughout the year, and hence it will be impossible to maintain a constant staff of workers; on the other hand, a judicious mixture of crops will keep a constant number of men fully employed for the year, and hence enable the farmer to keep good workers in employment. In deciding on a rotation, therefore, a farmer should bear this aspect of the question carefully in mind, and it may prove desirable to introduce one or more crops especially, to even out the labour demand. Mechanisation is particularly valuable when it can be applied at a season of peak labour demand; hence the rapid spread of combine harvesters for grain and the urgent need of a satisfactory harvester for the potato crop.

(iv) *Diversification*

No one likes to have all his eggs in one basket, and this is particularly true of the farmer, whose basket must necessarily be an unreliable receptacle. For cereals, sugar beet and potatoes minimum prices are fixed by the Government after annual Price Reviews and deficiency payments are made to bring average prices received by farmers up to these minima. But in some years the actual market price exceeds the guaranteed minimum for one or more of these commodities, to the producer's benefit, and in all years some farmers sell at prices well above the average, to their own profit. With crops other than those concerned in the Price Review price depends on the market at the end of the season of growth and this is hazardous in forecast at the time of sowing; in particular, the market garden type of crop, which is very expensive to grow, is subject to violent price fluctuations. Thus the farmer who grows a variety of high value crops is in a very much stronger position than the specialist, who relies too much on one or two such crops and who may be reduced to penury by low prices for his speciality in a few successive years.

A further cogent argument for diversity of cropping is provided by the vagaries of the climate. In some years wheat may do well, in others barley or roots, and few farmers have such a comfortable bank balance that they can afford to run the risk of too much reliance on one or two crops, which may all do badly in one year.

It is not difficult to see, therefore, that diversity of cropping and the inclusion of a large proportion of directly saleable crops are very desirable for purely economic reasons. This reinforces the arguments already advanced in favour of diversity, on the scores that it will do much to control diseases, pests and weeds, will tend to maintain the farm in good condition, and will create a level labour demand throughout the year.

(v) *Arranging the Sequence*

The mere growing of a multiplicity of crops does not, of course, ensure a sound rotation; having decided what crops to grow, the farmer has to arrange them in an orderly sequence. The considerations which have been dealt with already will determine roughly the acreage of each crop which will be required; the arrangement of the actual sequence is usually a simple matter, only demanding a knowledge of what is involved in the growing of the various crops. For instance, short leys take a full year to become established, and are therefore sown in spring under a corn crop, which latter pays the rent of the field during the unproductive year of establishment of the seeds. Wheat requires land in good heart and is therefore usually taken after seeds or beans, and frequently after potatoes. Wheat can be grown after a root crop which is carted off the field, though, in the case of sugar beet, folding of the tops with sheep may delay the clearance of the field too long for wheat; barley often follows sugar beet, therefore, and barley and oats are the usual crops to succeed fodder root crops which are folded. Roots and potatoes (heavily manured and thoroughly hoed), seeds and beans are all recuperative crops and should be spaced as equally in the rotation as possible, so that the exhaustive cereal crops may not all come together. There is, indeed, very little room for originality in arranging the actual sequence, and the practical details of crop husbandry usually dictate a sequence resembling one of the well-tried examples dealt with below.

(vi) *Relation to the Number of Arable Fields*

This point is of minor importance, but it is inconvenient and leads to wasted time and space to divide a field for the growing of different crops. Arable farmers should consider themselves lucky if they have big fields, and to split them up into strips of different crops leads to inefficiency and betokens a mind more suited to a smallholder. With ten arable fields of more or less equal size a five-course rotation would be preferable to a four-course, but such considerations must not be allowed to override the much more important ones discussed above.

Farming in a rotation necessitates the division of the arable land into sections, 'breaks' or 'shifts' of approximately equal area; it is common to speak of the part of the farm devoted to, say, roots, in a given year as the 'root break'. In making the division into breaks it is as well to take account of the fertility of the fields; thus if there are two fields of much lower inherent fertility than the rest of the farm they should be allotted to different breaks, so that they will not carry the same crop together. It is a convenience to arrange that blocks of adjacent fields are in the same breaks, as this facilitates supervision of labour and minimises transport of implements. Many farms include more than one soil type and consequently it is quite common to find two or even three rotations followed on different parts of the same holding. A number of rotations allow for alternatives, e.g. seeds or beans, mangolds or kale, and these alternatives should be taken on successive occasions; thus a four-course rotation including 'mangolds or kale', becomes an eight-course rotation for a particular field, in which mangolds are grown on the field the first time round and kale the second. 'Cross cropping' is a term used to denote departure from a rotation, as, for instance, is necessitated by the introduction of a new crop; in times of high prices it may be desirable to grow more corn for a few years, or a field that is reasonably clean may carry corn in place of a fallow, both these being examples of cross cropping. In farming circles cross cropping is almost a term of reproach, because in most instances it denotes exhaustive farming, as where several cereal crops are taken consecutively; in general it is done with the object of temporarily enriching the farmer at the expense of the land. Cross cropping must be carefully distinguished from catch cropping, which is dealt with below (pp. 35–43).

EXAMPLES

It is proposed to discuss the common rotations followed in Britain in order of their length from 1 year to 4 years, and from that point to deal with light and heavy land separately, because all the longer rotations can best be viewed as modifications of well-known four-year rotations.

New varieties of cereals which have been introduced during the last twenty years have made it difficult to be incisive in speaking of rotations. Previously there was no good yielding variety of spring wheat and to sow oats or barley in autumn was to run grave risk of the crop being killed by winter frost. Using appropriate varieties, each of these cereals may now be sown either in autumn or in spring with every prospect of success. It is still true, however, that wheat is relatively unsuitable for sowing immediately after any cereal crop, and danger of diseases and pests makes it better to vary the cereals in a rotation. Thus in what follows wheat will be taken as an autumn-sown crop with oats and barley as spring-sown, but it must be realised that in practice the three cereals are largely interchangeable without radical alteration of the rotation.

ONE-COURSE ROTATIONS

The term 'one-course rotation' is almost a contradiction, because the majority of cases that occur are cases of continuous cropping; there are, however, instances of one-course rotations which include more than one crop, and consequently the inconsistency may be pardoned. All the arguments advanced in favour of farming in a rotation can, of course, be adduced against one-course rotations, especially when they are definitely cases of the continuous growth of one crop.

Many attempts have been made to grow wheat continuously, and it must be confessed that in some instances the results have not been entirely disastrous. It is well known that virgin lands in other countries have been farmed in this way for years, but generally fallows have been introduced at fairly frequent intervals, and the farming has been extensive; that is, cheap land has made it possible to make a profit even though yield per acre has been, according to our standards, extremely low. Even in those circumstances the system bears no signs of permanency, the older settled lands being now necessarily farmed on a more reasonable system of cropping, the fertility accumulated over long centuries having been exhausted.

In this country the best known case of continuous wheat growing is on Broadbalk Field at Rothamsted, which has carried wheat since 1843; on this field the plots which annually receive generous dressings of fertiliser still give yields approximating to the average of the country. But this is a very special field, and it was only possible to keep it reasonably free from weeds under continuous wheat growing by dint of hand weeding on a scale only possible on an experimental field. At the present time it is not possible to continue with the requisite weeding even on that field, and it has been necessary to divide the field into five sections, running across the experimental plots, and to fallow one of the sections each year; in other words, the field has ceased to be a case of continuous wheat growing. Another much-quoted case is that of Mr Samuel Prout (and his son), of Sawbridgeworth in Hertfordshire, who is often cited as a successful farmer, who grew wheat continuously for upwards of 40 years, from 1861 onwards. Situated near London, in the days when the streets of that city rang to the sound of horses' hoofs, his method was to sell both the grain and the straw to the city, his waggons returning laden with horse manure. But examination of his cropping records reveals the fact that, though he drilled a large proportion of his farm with wheat, he grew other crops as well; though the evidence is that, with the copious supplies of horse dung, his soil did not deteriorate under the system, he grew seeds or beans or fallowed his fields about every sixth or seventh year, and sometimes more frequently, if the weediness of the fields showed it to be necessary. Under the Wheat Quota of 1931 a few farmers grew wheat for 2 years running on the same field and many, in patriotic zeal, did so during the Second World War. Good land that had lain in grass and been well grazed for a long period produced bumper crops of wheat for 2 or 3 years and in a few cases for as much as 6 years in succession. But there is no lesson for the ordinary farmer in these very special cases. Wheat is an exhausting crop, which requires land in good heart, whilst if grown continuously, or too frequently, disease often becomes serious, and, most important of all, the land becomes very foul with weeds.

Barley, with its shallow rooting habit, is even more unsuitable for continuous growing than wheat, but it is taken for 2 or even 3 years successively with some measure of success; this, however, is not generally a one-course, but part of a longer, rotation.

Some farmers grew sugar beet continuously after the Sugar Beet Subsidy was given in 1923, but, as mentioned on p. 7, the practice is now prohibited by the terms of the contract with the factory, because of the grave danger from eelworm. The same danger, but presumably to a less degree because fewer roots are grown per acre, attends the continuous growing of mangolds, but this has been by no means uncommon in the past. Mangolds involve much heavy carting, both of farmyard manure to the field and of roots home, and therefore it is very convenient to grow them on a field near the buildings; consequently some farmers have grown them continuously on a handy field, and it seems that good crops can be grown over a long series of years, although the danger from pests must be greatly increased.

Even potatoes have been grown year after year on the same field (light fields suitable for earlies) in some districts, but the danger from eelworm concentration is a very real one, and has manifested itself calamitously in many cases. In general this is not an instance of continuous cropping, but of a one-year rotation, because early potatoes are dug soon enough to permit the growth of a second crop in the same season. In the Ayrshire early-potato district the common practice is to sow Italian rye grass for sheeping or green manuring in the autumn, whilst in Bedfordshire it is common to sow white turnips after early potatoes, the turnips being sold as a vegetable, or ploughed in during the autumn. In both these cases the second crop is valuable in that it utilises the residues of the heavy nitrogenous manuring given to the potatoes, and thus prevents leaching in the late summer and autumn. An alternative method is to interplant the early potatoes with brussels sprouts, which are left established when the potatoes are dug; this practice, however, is almost confined to private gardens and is rarely seen on a field scale.

TWO-COURSE ROTATIONS

The best known two-course rotation is that used in so-called 'dry farming', in large areas of North America and Australia, where the rainfall is barely sufficient for wheat; it is

Bare fallow,
Wheat.

The principle underlying this rotation is that 2 years' rainfall is accumulated for the growth of one crop, for land which is fallow loses less water through evaporation than land on which plants are growing and transpiring water. Canadian scientists have found that 25 per cent of the annual rainfall is conserved but studies in South Australia have shown that the water accumulated in the fallow is only equivalent to 0·6 in. of rain. The latter studies showed that the main benefit derived by the wheat from the fallow was in the amount of available nitrogen in the soil, as during the fallow bacteria experience favourable conditions for converting nitrogen from the unavailable to the nitrate form; the amount of nitrate nitrogen present in the soil after a fallow was shown to be nearly double the amount present after a crop. It would be idle to deny the importance of the above rotation in the countries mentioned, but in this country conditions are very different, because the rainfall is sufficiently heavy—over most of the country, too heavy—for wheat, and the rotation consequently has no place in British farming. Mr George Baylis, who farmed successfully on a very large scale in Berkshire from 1866 onwards, used, however, a sequence of cropping similar to the above. His rotation was corn-fallow, except that clover occupied every third fallow break. His corn consisted largely of winter barley, though wheat played an important part in his system.

A fairly common two-course rotation in British agriculture is

Roots,
Corn.

On light-land farms situated near a sugar beet factory this rotation is moderately suitable. The root break would be nearly, if not quite, all sugar beet and the corn break largely barley, though wheat might follow the earliest cleared beet. There is no difficulty in criticising this rotation, with its great dependence on two crops, its uneven labour demand and its shortness inviting trouble from disease, but all these considerations may be held to be overridden by the vicinity of the factory, which makes sugar beet a pre-eminently suitable crop. Some farmers in the Fens devote half their acreage to potatoes, and consequently farm on what is essentially the above rotation, their system also being open to the criticisms just given. In Bedfordshire, too, a similar rotation is found on light land suited to potatoes. In that district the corn is nearly all barley and the root break carries market-garden crops and potatoes successively, so that the rotation really becomes a four-course rotation of roots-barley-potatoes-barley. It is a common practice to undersow the first barley with red clover, which is allowed to grow up after harvest, so that there is a fair growth to plough in in the late autumn, this being regarded as a very useful preparation for the potato crop.

Farmers with only a small proportion of light land not infrequently follow a two-course shift, of roots followed by corn, on their lighter fields. Since the rotation is only used on a small proportion of the farm the main arguments against such a short rotation do not apply in that case, whilst disease can be avoided by varying the root crops, and also the corn crops, grown on the same field in successive breaks.

THREE-COURSE ROTATIONS

In the history of British agriculture probably the most important rotation has been

Autumn corn,
Spring corn,
Fallow.

This is the rotation on which much of the country was farmed under the old manorial system; it was introduced about the time of the Roman occupation and persisted throughout the feudal and mediaeval epochs, only being gradually displaced during the seventeenth and eighteenth centuries. A full description of the system may be found in text-books on agricultural history, and there is one village—Laxton in Nottinghamshire—which still has its manorial court and the system in operation; this, however, can only be regarded as an interesting historical survival. The autumn corn consisted of wheat or rye, and the spring corn of barley or oats (occasionally peas or beans), and after harvest stubbles were run over by sheep; the fallow was sheeped during the spring and early summer, being broken up subsequently. The obvious weakness of the rotation lay in its waste of one year in three, but the main weaknesses were inherent in the system, which required that each man farmed several isolated strips in the three large fields; this led to waste of effort, and the fact that innovations were usually ruled out by the manorial court precluded progress. It is probably true that the rotation led to a slow exhaustion of the land, but the deterioration must have been extremely slow, since the system held its ground unchallenged for upwards of a thousand years; in fact, it was only the introduction of roots and clover about A.D. 1700 and the subsequent enclosure movement which eventually sounded its death knell. A very similar rotation is still widely practised in parts of eastern Europe, and it is a striking fact that when the combine harvester was first introduced into this country it was this rotation which was commonly used by combine pioneers; but its complete reliance on corn, its waste of one-third of the arable land, its tendency to make the land foul and to encourage disease have proved it to be inferior to rotations which include roots and leguminous crops, even on fully mechanised farms.

It is now a very general practice to take two corn crops in succession, following them with a recuperative crop such as beans, seeds or roots, but actually the sequences are extended to six-course rotations and so they will be discussed later. On very rich land, suited to the growing of high-value crops, rotations are usually the reverse of the above, in that they consist of one corn crop followed by two root, or similar, crops; thus on the fens and silts of Cambridgeshire and Lincolnshire the most common rotation is

Wheat,
Potatoes,
Sugar beet.

In the farming of these rich soils livestock play only a small part, so that crops that are directly cashable occupy practically all the farm; fallows are unnecessary in such circumstances, the opportunities for cleaning being clearly adequate, and the high rent of the land militating against the waste of a year. The farming is very intensive and the relatively low-valued wheat only finds a place because straw is required for covering potato clamps and for such stock (pigs and perhaps fattening cattle, in addition to horses) as are carried. It might be thought a weakness in this rotation that wheat follows sugar beet, but on that land this presents no difficulty; in the first place the sugar beet tops are almost invariably ploughed in, so that the fields are quickly available after the lifting campaign, and in the second place wheat is best if not sown too early owing to the danger of too much growth in the autumn. Some fen farmers do, however, reverse the order of the last two crops in the rotation, so that wheat follows potatoes, which should be cleared well before the end of October, leaving adequate time for the sowing of the cereal. A large acreage of potatoes and sugar beet creates an uneven labour demand, much being required in spring and early summer and also in early autumn; the provision of labour is, in fact, one of the fen farmer's greatest crosses, but he overcomes the difficulty by encouraging long hours by means of piece-work payment, and by employing women. The rotation appears to be lacking in variety, and it is true that a very large proportion of the income of the farmers is obtained from potatoes and sugar beet. But the rotation is not adhered to at all rigidly; many fen farmers grow celery, carrots, green peas, mustard seed, etc., on suitable fields, whilst the corn is not exclusively wheat, some oats being grown and occasionally barley. One-year leys are also taken, commonly of Italian rye grass on the fens but generally of broad red clover on the silts.

FOUR-COURSE ROTATIONS AND THEIR DERIVATIVES

(i) *Light Land*

During the seventeenth century clover was introduced to this country and turnips gradually passed from garden to field culture; at the close of the century Jethro Tull introduced the drill and horse hoeing, and thus enabled turnips to take their place as a fallow crop. It was Townsend, however, who developed the full possibilities of these crops and evolved, on his own estate, the Norfolk four-course rotation, on which a very high proportion of the lighter lands of this country was farmed for nearly 200 years; that rotation is

Roots,
Barley,
Seeds,
Wheat.

This was more than a rotation, because it carried with it a whole system of farming. The root crops were largely turnips or swedes, which were folded by sheep during the winter; these sheep folded the seeds leys during the summer and so the rotation carried with it, as a necessary concomitant, an arable sheep flock. Further stock were required to tread down the straw into farmyard manure, and the traditional policy was to fatten bullocks in yards during winter, these bullocks also consuming large quantities of roots. Thus the system produced wheat, barley (high-quality barley for malting), mutton and lamb, wool and beef, and on it the light-land farmer was very successful.

In the circumstances obtaining in Townsend's time, and in fact until the war of 1914, this rotation was nearly perfect. There is no doubt that under it the fertility of many thousands of acres was not merely maintained but improved; this, it must be appreciated, occurred before chemical manures became obtainable, and even now little expenditure on that head (except for phosphate) would be warranted. The rotation included two shallow-rooted crops (turnips and barley) and two deep-rooted ones, but it effected its improvement in fertility mainly by means of the heavy stocking which traditionally accompanied it. The frequent folding of the land with sheep played an important part in this, but so also did the rich cake-fed dung provided by the yards of fattening cattle. The rotation, with its varied crops and the frequency of roots, also led to clean farming. As regards disease very little trouble was experienced; on land with a low lime reserve there was some danger of finger-and-toe disease on turnips, but this could be avoided by alternating with mangolds, as could clover sickness by alternating the seeds ley with beans or peas. Finally the rotation created a very even labour demand, the only time of the year when extra help might be needed being the summer, when root hoeing called for much labour; this was aggravated by the fact that root hoeing overlapped hay time. Before turning to the shortcomings of the system under present-day conditions it must be emphasised that under it very poor light land was kept in cultivation and produced reasonable crops. As the system declined much of the poorest land on which it had been successfully practised simply went out of cultivation. Afforestation has proved a useful alternative for some of this land but Lord Iveagh has shown by his magnificent pioneering work in Suffolk that it can be made, and kept, highly productive by a system of dairy farming with lucerne/cocksfoot leys.

It must be agreed that the Norfolk four-course rotation has much to recommend it, but during recent years it has ceased to hold the prominent place it occupied for so long. The reason for this is that the price of labour, taking hours as well as wages into consideration, has risen more than ten-fold during the last four decades, and folding roots have become very expensive to grow; there has been no more marked change in agriculture in this period than the decline in the numbers of arable sheep. It is not only expensive to grow roots but also to feed them, and this has militated against arable sheep, and therefore against this rotation; furthermore, sugar beet has been profitable and the farmer has therefore been anxious to introduce it into his system. There are now very few, if any, farms in the country where the old Norfolk system is fully practised. It persisted longest (until 1940) on the Wolds of the East Riding of Yorkshire and there it was military, not economic, necessity that sounded its death knell. Most of the area still devoted to it was scheduled as a training area, the land still being farmed though troops had full training use of it, paying compensation for the damage they did. In these circumstances it was impossible to keep an arable flock and the area was denuded of sheep; with the release of the land the farmers were wisely hesitant to return to their previous system. The present position is that financial success would necessitate great technical efficiency and the possession of a well-known flock from which ram lambs could be sold at high prices for crossing on grassland ewes; no one but an outstanding sheep master and first-rate dealer in cattle could hope to make ends meet if he kept strictly to the Norfolk four-course rotation and the stately system which tradition associated with it. Consequently many modifications of the Norfolk four-course rotation have been introduced; the general trends have been to a reduction, or the elimination, of the proportion of fodder roots, and an increase in the proportion of crops which are directly saleable. Thus it is mainly as a generator of others that the Norfolk four-course rotation is important at the present time; it is true to say that practically all common rotations now to be found on light land have been derived from, and can readily be recognised as modifications of, the old Norfolk sequence. The following are the more important modifications to be found in practice to-day.

(*a*) *Substitution, in part or in whole, of sugar beet for the fodder roots.* The effect of this is to increase the proportion of directly saleable crops from one-half to three-quarters, substitution introducing a particularly high-value crop. On farms which include some moderately heavy land a common procedure has been to lay that land down to grass and to run a flock of grass, instead of arable, sheep. On farms unsuited to grass the arable flock may not have disappeared entirely, but may have been reduced to conform to the fact that sugar-beet tops have only approximately half the feeding value, acre for acre, that fodder roots have. In some cases sheep have been displaced altogether, and then the farmer has greater freedom with his leguminous break; he may take trefoil or sainfoin for seed, or grow peas in place of clover.

(*b*) *The lengthening of the rotation.* It has been said that one of the drawbacks of the Norfolk four-course rotation was the small proportion (one-half) of the crops that were directly saleable; the simplest, and a very common, modification, has been the addition of another crop, to give

Roots,
Barley,
Seeds,
Wheat,
Oats or barley.

This need not disturb the general structure of the farming system appreciably, and it performs the useful functions of raising the saleable crops to three-fifths of the area, and of reducing the fodder roots to one-fifth; some of the root break is usually devoted to sugar beet, giving a further increase in the saleable crops, and necessitating a greater reduction in the number of sheep.

The above change consists of doubling the second corn crop of the old Norfolk four-course rotation; many farmers have gone a step further and have doubled both corn crops to give

Roots,
Barley,
Barley,
Seeds,
Wheat,
Oats or barley,

and sometimes, yet another corn crop is taken at the end to make a seven-course rotation. On farms with a high reputation for first-class barley, this rotation has much to recommend it, and it is a very common rotation on light chalky land; the farming system must clearly depend on corn prices for its success, and there is an undoubted danger of loss through take-all on the wheat and barley. In some cases the sheep flock is retained, with reduced numbers, and then the root break consists chiefly of kale; often a dairy herd has taken the place of the sheep flock, kale being folded by the cows during autumn and early winter. In a few cases a clean sweep has been made of livestock, and then the root break consists of either sugar beet or bare fallow with mustard grown in late summer for green manuring; the seeds leys are then often trefoil or sainfoin cut for seed. The traditional farmer frowns on this long rotation with its great reliance on corn, particularly as, where it is adopted, the livestock carried by the farm are reduced to a very low level; since 1920 the numbers of cattle fattened in yards have fallen markedly and on some farms straw stacks have been allowed to accumulate or have been sold if the market price was good. The land has, therefore, been deprived of its great sources of fertility—sheep-folding and farmyard manure—and it is interesting to speculate on the power of green manuring and artificial fertilisers to take their place.

(*c*) *The introduction of potatoes.* On deep light land which does not contain much chalk, the potato crop has been introduced by many farmers. In some cases it has merely been inserted between the seeds and the wheat of the Norfolk four-course rotation; this has the advantages that the seeds leys may be grazed late in the autumn, that potatoes do well on land containing much humus, and that potatoes leave the land well for wheat. The best-known modification of the Norfolk rotation to include potatoes is, however, the East Lothian one:

Roots,
Barley,
Seeds,
Oats,
Potatoes,
Wheat.

This is an excellent rotation with a high proportion of saleable crops, a good variety and an even labour demand. It is possible to permute the last three crops in several ways: wheat-potatoes-oats, potatoes-wheat-oats are sequences often found in practice.

(*d*) *Extension of the seeds ley*. This most important and widespread modification is dealt with below in the section on 'Alternate Husbandry'.

(ii) *Heavy Land*

There is no widely followed four-course rotation on heavy land corresponding to the light-land Norfolk one, but generally its place may be said to be taken by

Bare fallow,
Wheat,
Beans,
Wheat.

There are very few farms on which this rotation is adhered to strictly, but the general structure can be seen running through the crop sequences on much heavy land. A bare fallow is to be avoided if possible, because of the waste of the land, but on very sticky soil it may be the only satisfactory way of controlling perennial weeds; rarely is a whole break devoted to it, the cleaner fields, or parts of fields, usually carrying mangolds or kale. Beans are very suitable for such land, but probably a part of their break would be taken by red clover, or by oats and tares for hay, or by a crop for silage, though the last is not so common as it ought to be. It follows that in practice this rotation generally becomes an eight- or sixteen-course one, with wheat occupying about half of the arable acreage; wheat, it should be remembered, is one of the few crops of reasonable value suited to clay land, and necessarily the saleable crops are almost restricted to it. Farms on which this type of rotation is practised normally consist largely of permanent pasture, and the sheep are of the grass breeds, and rarely find themselves within hurdles; dairy cows and store cattle play a large part in the economy of the farms, and consequently there is no difficulty in turning the straw into farmyard manure.

On heavy land, as on light land, the modern tendency is to lengthen the rotation, and a common modification of the above rotation is to take oats or barley after the second wheat, thus raising the acreage of corn crops to three-fifths of the arable land. The analogy with light land is most clear in the case of those clay-land farmers who have extended their rotation to a six-course one, which is of the form

Beans,
Wheat,
Oats or barley,
Seeds,
Wheat,
Barley or oats.

The tendency, it will be observed, both on heavy land and on light land, is to work to a double three-course shift, with a recuperative crop every third year. The above rotation does not include a bare fallow at all, which is an advantage if it is found possible to keep the land clean without one; if couch and water-grass become troublesome there is a chance for taking a bastard fallow in the seeds break, the ley being broken up after a hay crop has been removed. The rotation may be criticised on the score that it involves an uneven labour distribution, since the crops consist almost entirely of corn, and largely of autumn-sown corn. From 1930 to 1939 many farmers working on this rotation were tempted to displace a large proportion of the oats and barley with wheat, and it was not difficult to find fields which grew this crop for 3 or even 4 years out of 6. A farmer is usually rather ashamed of such a procedure, but when wheat was the only cereal even to promise a profit his lapses were at least excusable; all too frequently they carried with them the dire penalty of poor yield. The above six-course rotation is probably a suitable one for the general farmer on heavy land in the south and east of the country to-day, but it does require good management to pursue it successfully; with two-thirds of the land under corn, every opportunity must be taken of cleaning the land, and, in particular, the quick breaking of stubbles after harvest is very desirable. Some farmers cling to their principles and refuse to grow two white-straw crops successively, and they use a rotation of the following form:

Fallow or roots,
Oats,
Beans,
Wheat,
Seeds,
Wheat.

With this rotation it is much easier to keep the land clean, but failure of the seeds is not uncommon. The fungus causing clover sickness sometimes attacks beans, and hence growing seeds and beans too close in the rotation is dangerous.

ALTERNATE HUSBANDRY

The traditional practice in the drier parts of the country has been for some fields of a farm to be permanently arable, farmed on one of the rotations already discussed, whilst the other fields are permanently grassland. In the wetter districts, on the other hand, rotations have become established in which the seeds ley is extended to 3 or more (rarely over 7) years and the whole of the farm is included in the rotation, except possibly an inaccessible, inconvenient or hilly field. In this system each field has periods of years under arable crops alternating with periods under long ley and hence it is known as alternate husbandry or the long ley system. There is nothing rigid in the system, indeed its flexibility is one of its great merits. Some farmers shirk ploughing up leys until they show signs of deterioration, whereas others aim to keep the ley and the arable part of the rotation of equal lengths and consider that the amelioration of the soil by the ploughing in of the ley only lasts for 3 or 4 years. It should, perhaps, be pointed out that in long leys the plants used are not the same as in short leys; in the former case the foundation of the mixture is usually provided by perennial rye grass, cocksfoot or timothy and wild white clover whilst the most important constituent of the latter is red clover.

There are several powerful arguments which can be advanced in favour of the long ley system. Whilst a field is in grass, and properly grazed, it accumulates fertility. The dung and urine of grazing animals and the roots of the clovers add to the reserves of plant food and when the grass turf is ploughed up the incorporation of organic matter with the soil greatly improves physical condition. The accumulated fertility is cashed out in the arable crops which follow the sward, heavy crops being grown with moderate expenditure on chemical fertilisers. It is not only the arable crops which benefit. New grass is more productive than old and so the system leads to improvement both in arable crops and in pasture. Further important points in favour of the system are the controls it gives of plant, and, even more, of animal diseases and of weeds.

There has never been any difficulty about alternate husbandry in districts of high rainfall; leys are easy to establish and are highly productive throughout their life. Thus it is that the system is of respectable antiquity in most of Scotland, Wales and Ireland and on the western side of England. Records show its adoption on Exmoor 150 years ago, and the successful practice and the writings of Elliot of Clifton Park in Roxburghshire in the last years of the nineteenth century did much to extend the system's adoption. In these wetter districts wheat and barley are rarely successful, the only reliable corn crop being oats; of the root crops, turnips and swedes are the most common, although kale has increased its area in recent years. Consequently, a very common form of rotation, to be found over wide expanses of the British Isles, is:

Turnips,
Oats,
Seeds,
Seeds,
Seeds,
Seeds,
Oats.

This form is palpably a simple adaptation of the Norfolk four-course rotation but there are many variations in detail to be found within the general form. The length of the seeds ley varies considerably from district to district. In Aberdeenshire farmers have two well-known variations—the 'easy seven', which is the rotation set out above, and the 'hard seven' in which the ley is of only 3 years duration and is followed by two corn crops making four arable years out of the seven. In other districts, Northern Ireland for instance, leys are down for 7 years with successions of three arable crops between them. The flexibility of the system should be carefully noted. In times of depression commitments can be reduced by allowing the leys to remain longer, whilst in war-time they can be ploughed up earlier and this, coupled with the doubling of the corn crops, rapidly swings a farm over to a high proportion of arable crops; this causes no dislocation of the farming system and is in sharp contrast with the revolution brought to other farms where large acreages of permanent grass have to be ploughed up. Where land is farmed highly on the alternate husbandry system, and clover-rich leys are heavily grazed, the fields accumulate too much fertility for the oat crop that follows the ley; to avoid lodging of these crops it is becoming more common to take roots first after the ley, following them by two corn crops.

The reason why alternate husbandry has not been common in districts of low rainfall is that until comparatively recently leys, in those conditions, have been very unproductive in their 2nd, 3rd and 4th years. The inclusion of bulky but short-lived plants in the seeds mixture could provide luxuriant growth in the 1st year of the pasture but then these species died out and it took another 2 or 3 years for the longer-lived species (wild white clover in particular) to become established. But the work done at Cockle Park in Northumberland showed that poor pasture on heavy clay land might be improved as much as sixfold by the application of basic slag and that the improvement was due to an astonishing increase in the wild white clover content of the herbage. As a result, the supply of wild white clover seed improved rapidly and between the two World Wars this seed became reasonably plentiful, though it remains expensive. There must always be some risk of failure in establishing a long ley in a dry district but with wild white clover included in the seeds mixture, with proper manuring and with some cultural care the risk is not unduly high; in recent years the introduction of S.100 white clover has further improved the position, so that now a long ley can be very highly productive in each of its first 4 years. Between 1925 and 1939 some farmers in the south and east of England introduced long leys into their rotations with marked success and the movement gained impetus rapidly during the war of 1939–45. Large acreages of corn were required and, to avoid exhaustion of the arable land, County Agricultural Executive Committees encouraged, indeed ordered, farmers to rest fields by putting them down to ley, corn acreage being maintained by the further ploughing up of old pastures. This had the double effect of keeping up the yield per acre of corn crops and of providing better pastures, so that a greater head of cattle could be maintained on a much reduced acreage of grassland. Under the stimulus of war and with reasonable profit-margins the policy of 'taking the plough round the farm' made great strides, but it takes a long time to adapt an arable area of big open fields to alternate husbandry. The cost of fencing and laying on of water may be prohibitive. Electric fencing, government grants for water supplies and hay leys based on lucerne may all do something to meet the situation, but even the farmer who is fully persuaded of the wisdom of the step may well take a decade over the conversion. The fear is that difficult economic conditions may incline farmers to return to the easier and cheaper system of permanent pastures; this would be regrettable because alternate husbandry is a high type of farming in which very productive pastures prepare for bumper arable crops.

As yet there are no long ley rotations for dry districts which have been proved by long practice. On heavy clay land a suitable sequence is:

Seeds,
Seeds,
Seeds,
Seeds,
Wheat,
Beans,
Wheat,
Oats or barley.

As there must always be some element of risk in establishing a long ley where rainfall is low it may be that there will be a tendency to allow a successful ley to occupy the land longer, possibly for 6 or more years. If the proportion of land in tillage is to be kept up this will mean longer sequences of arable crops; these sequences must be very much the same as the rotations discussed earlier in this chapter for the various soil types, and will include roots and, presumably, one-year leys. After a three- or four-year ley it may be possible to grow three useful cereal crops and then return to ley but generally no more than two cereals should be taken successively. A pulse, a root crop or a short ley should come after that and then two more cereal crops will usually succeed. Hitherto, the last of the arable series of crops in an alternate husbandry rotation has almost universally been a corn crop, which acts as a nurse for the young seeds. But with low rainfall, leys are much more certainly established when the seeds are sown on bare ground in early spring, preferably in March. After a corn crop the stubble can be ploughed and the furrows left to weather through the winter, but if the last arable crop is roots the land should be cleaner for the small seeds. March sowing should produce abundant grazing by June, and the only losses as compared to sowing under corn in the previous spring are a problematical autumn grazing of the stubble and a May grazing which can usually be spared. Where land is foul with perennial weeds it may be better to fallow the field and sow the small seeds as the rain comes in July or August.

CATCH CROPPING

A catch crop is one that is stolen between two main crops, which leave the ground unoccupied for a considerable time. For instance, where roots follow corn, the latter is harvested about August and the roots are not drilled till the following spring or summer; there is consequently sufficient time for another crop to be grown in the interval, without curtailing the growing periods of the two main crops. This is a simple, clear-cut case, but it will be difficult to avoid inconsistency in speaking of catch crops, because it is often hard to decide which is the catch, and which the main, crop.

Different catch crops are grown for a variety of reasons, but the following may be taken as the main advantages gained by growing them:

(*a*) *Helping to pay the rent.* Catch cropping is an intensive form of farming and clearly makes for a fuller utilisation of the land.

(*b*) *Replacing a main crop which has failed.* Probably the most frequent occasion for a catch crop under this head is when a root crop fails, as happens all too commonly in a dry season; it is possible to fill in the gap created in the food supply for arable sheep, by sowing rape or white turnips late in the summer. The seeds ley is also apt to fail, again particularly in a dry summer, and the failure is revealed when the corn crop, in which it is sown, is harvested; it is not too late to sow crimson clover in September, or, alternatively, the field may be broken up and oats and tares sown to provide a hay crop to compensate for what has been lost.

(*c*) *Providing food at a time when it is particularly needed.* The great example of the usefulness of a catch crop in this respect is provided by the arable-sheep farm, on which the flock spends the winter on roots and the summer on leys; root crops are often finished by the end of March, and, furthermore, the following barley should be sown by then, whilst leys consisting of red clover or sainfoin will not provide keep until well into May. It was, therefore, a common practice under the Norfolk four-course rotation to undersow the wheat with a mixture of trefoil and Italian rye grass, which will grow luxuriantly in April, and so fill the hungry gap which is liable to exist. It will be realised that it was not the practice to undersow all of the wheat with the mixture, but only such portion of it, usually one field, as the farmer estimated would be needed to carry the flock over from the roots to the normal seeds leys; having provided the necessary food for this interval the field was broken up, and took its normal place as part of the root break. Probably an even more common catch crop in this situation was rye which can be sown after the corn harvest and fed off earlier than trefoil and rye grass; this crop provided valuable and very early spring keep, without prejudicing either of the main crops.

The opportunity for this type of catch crop has shifted from the arable flock to the dairy herd, folding of which is facilitated by the electric fence. With cows there may be a very expensive hungry gap between the exhaustion of the winter foods (roots and silage and, in desperate circumstances, the hay as well) and any considerable growth of grass. Rye and other cereals sown in the previous September can fill this gap and the field can be ploughed in ample time for kale. Italian ryegrass, undersown without any clover in the previous cereal crop and heavily manured with nitrogen, is becoming increasingly popular for the March/April period. With continued generous manuring it may maintain itself for several grazings and occupy the land for one or even two seasons, but then it ceases to be a catch crop and becomes a ley.

(*d*) *Preventing the leaching of nitrogen from the soil.* During late summer, when the soil is warm, nitrogen is rapidly converted by soil organisms into the nitrate form, but if the land is fallow, or if it carries a corn crop which will then be ripening off, the nitrate will not be taken up; it is therefore very liable, at least on light soils, to be washed through the soil by the rains of early autumn. A catch crop may be sown to take up the nitrogen, which is later returned to the soil in a less available form, either by ploughing in, or folding off, the catch crop.

(*e*) *Recuperating the land.* Catch crops are sometimes grown mainly for the purpose of green manuring, and in other cases they are fed on the land, which benefits thereby considerably. In addition a catch crop may be valuable in smothering weeds.

(*f*) *Saving a bare fallow.* A seeds ley cut for hay in June may be left to provide a second growth; another alternative is to sacrifice the second growth and to break the land up for a bastard fallow, as may also be done after a silage crop. In these cases the crops may be regarded as catch crops, though it is rather the bastard fallow which is stolen.

It will be seen that there is a formidable list of benefits, one or more of which may be gained by catch cropping, but there are some disadvantages associated with the practice. The chief one is the danger that the catch crop may spoil the main crop which succeeds it. Thus a catch crop grown before roots may utilise moisture, the supply of which, on light land, is often the chief factor determining the success of the root crop; if, also, the catch crop is a good one and carries the flock for some time, there is the temptation to let it occupy the ground for too long, leaving inadequate time for the preparation of the ground for the roots.

If a seeds ley preceding a bastard fallow be regarded as a catch crop, there must be set against the advantage of having a crop the fact that a bastard fallow does not give so good a chance of cleaning the field as a bare fallow, unless the latter half of the summer be hot and dry. The very fact that a catch crop is often sown with little preparation, and rarely receives any manure, may be cited against it; often it is grown too cheaply, and is therefore a failure, so that what time and money have been expended upon it are wasted.

It is very difficult to classify catch crops, but they may be divided roughly into two groups. The first group is grown between a corn and a root crop, and may be designated winter catch crops; the second group is grown after a crop (or a fallow) which is finished by midsummer, or soon after, and may be called summer catch crops. The former is probably the more important group.

WINTER CATCH CROPS

These are grown chiefly for the reason already mentioned, that is, to carry an arable-land sheep flock or a dairy herd over the spring gap between root crops and leys. The value of the catch crop is really much greater than might be thought from the small proportion of the year over which it provides feed. With arable sheep the important point is that by filling the gap it enables a farm without grassland to maintain a flock continuously throughout the year. With dairy herds, for which this type of catch crop is becoming more common, the great gains are the shortening of expensive winter feeding and avoiding punishment of leys before they have got reasonable growth on them. Hence the aim is food as early as possible in spring and for this cereals sown very early in autumn and ryegrass sown under the previous year's corn crop are most suitable. The scheme shown on p. 39 shows the chief crops which may be grown, the catch crops being arranged roughly in order of earliness of feeding, and the following main crops approximately in order of sowing date.

The dual position of white turnips is explained by the fact that if this crop is sown in July it produces bulbs for feeding in autumn, and in that case is not winter hardy, whilst if it is sown after harvest it will form few and small bulbs, and the green tops will survive the winter frosts. Kale is shown as an early main crop, but it is often sown much later and might, with equal justification, have been placed lower down the list. In any case the second and third columns of the scheme below must not be regarded as tallying exactly; practice varies considerably with regard to sowing dates, and the earliness of the spring must clearly largely determine the date when the catch crop is cleared, as also must the area and yield of that crop. In general there will be time to sow mangolds, kale and maize after any of the winter cereals, but the catch crops lower in the list must be followed by a main crop which is also placed lower down in that column.

Preceding main crop	Catch crop	Following main crop
Corn	White turnips (tops)	Mangolds
	Winter rye	Kale
	Winter wheat	
	Winter barley	Maize
	Winter oats	Swedes
	Trefoil } alone or	Yellow turnips
	Italian rye grass } mixed	
	Crimson clover	White turnips (bulbs)
	Winter oats and tares	Rape
	Spring oats and tares	Mustard

It must be realised that these catch crops are often grown spasmodically by a farmer, as when he foresees a lean period for his stock, and are only sown on small areas. There is, however, one very well-known rotation which includes them in a regular manner. This is the Wiltshire eight-course rotation, on which a large area of light chalky land in that county and its neighbourhood was farmed with almost religious faithfulness until the war of 1914–18. The rotation was designed to provide the maximum amount of food for arable sheep and was

Catch crop and late roots,
Early roots,
Wheat,
Barley,
Catch crop and late roots,
Barley,
Seeds,
Wheat.

This rotation undoubtedly maintained the land in a high state of fertility and cleanliness, but the fodder-root proportion was even higher than in the Norfolk four-course rotation (i.e. three-eighths, cf. one-quarter); the large increase in the price of labour has, therefore, hit the traditional Wiltshire farmer worse than his Norfolk confrere, and consequently this rotation is not now practised. The late roots consisted of kale, swedes and turnips, whilst the early roots were usually mangolds or kale. The late roots shown in the first break were the last of the winter forage to be consumed, and therefore left the land too late for corn; hence another root crop was grown, and this succession of 2 years of folding crops made the land too rich for barley, so that wheat was grown next to 'steady the land down'. The system of farming built up on this rotation was a very fine one, and produced a large and varied amount of human food from land of inherently low fertility; to many it is a source of regret that economic changes have rendered the system unprofitable. A weakness of the system was that it tended to make the land 'sheep sick', as fields were folded six or more times in 8 years.

The crop selected for growing as a winter catch crop must be possessed of certain characteristics. It must of course be winter hardy, and it must grow early and quickly in the spring. It is important that it should produce a large bulk of nutritious fodder, and in this respect the crops shown vary considerably. Rye produces much bulk compared to the other cereals, but it must be consumed early because otherwise it becomes fibrous; crimson clover, if successfully established, produces a large bulk, but it, too, must be consumed in the young state, as later it becomes very tough and unpalatable.

It is not all farms that are suited to the growth of winter catch crops. The soil must be light and easy working, so that seedbeds may be quickly prepared; only light soils will warm up quickly in spring to give early growth, and only on them will early folding be possible. The climate of the locality is also important. What is required is a mild open winter, so that growth may be continued late in autumn and may be resumed early in spring; an adequate rainfall is very desirable and the growth of these crops is most common where it reaches 30 in. or more annually. Nevertheless, the chief consideration is the lightness of the soil, and so the winter catch crops are often to be found on light land in East Anglia, despite the severe winters often experienced in that part of the country, and despite a low rainfall: it is worth noting, however, that these crops have been most fully incorporated into a farming system under the equable climatic conditions of the south-west.

SUMMER CATCH CROPS

There are three quite common and distinct cases in which opportunity is presented for summer catch crops; these are following early potatoes, after a fallow and after a crop cleared about midsummer. The first case has been mentioned when single-course rotations were discussed, and it was seen that early potatoes might be followed by Italian rye grass (mustard or rape is used occasionally) or by white turnips. The catch crop requires little work or expenditure, as when the potatoes are lifted the land is left in a loose friable condition, and the next crop may be drilled forthwith; furthermore, the land is in a rich state, having been liberally manured for the potatoes, and, in fact, one of the main reasons for sowing a catch crop is to prevent the leaching of the surplus nitrogen from the soil.

Practically the only catch crop sown after a fallow is mustard, which is broadcast or drilled in July or August. It has three important justifications. First, as a nitrate catcher to prevent the loss of nitrogen rendered available during the fallow; secondly, for green manuring, the usual practice being to plough the crop in during the autumn; thirdly, to cover the ground during August, and so to act as a protection against the wheat bulb fly, which lays its eggs on bare ground at that time. In regard to the last it must be remembered that wheat is the most common crop to follow a fallow. In some parts of the country yet a fourth object is served by sowing mustard at this time; it will provide cover for game, and on land where the partridge shooting is valuable farmers often have an understanding with the shooting tenant that a certain acreage of it shall be grown and not ploughed in before a specified date.

There are only a few crops which are cleared early enough to allow time for a bastard fallow; the three chief ones are silage mixtures, oats and tares for hay and a seeds ley, and in the third case the fallow can only be made at the expense of the second growth of seeds.

In Table II an attempt is made to summarise the basic information required about the common catch crops; it must be realised that with these crops practices are far from standardised, so that the seeding times and rates given must only be taken as rough averages of the general practice. Some of the crops included are often grown as main crops, but all of them may be stolen between the normal crops of a rotation, and the details given only apply to the latter case.

TABLE II

Crop	Date sown	Seed rate	Date of feeding
Rye	September	4 bushels	March–April
Barley	September–October	4 bushels	April
Oats	September–October	4 bushels	April–May
Tares	October	2 bushels	May–June
	March	2 bushels	July–August
Oats and tares	October	3 bushels oats	May
	March	1½ bushels tares	July
Crimson clover	September	20 lb.	May
Trefoil	April*	14 lb.	April–May
Italian rye grass	April*	25 lb.	April–May
Italian rye grass and trefoil	April*	20 lb. Italian rye grass, 10 lb. trefoil	April–May
Mustard	July–August	14 lb.	September–October
Rape	July	6 lb	October onwards
White turnips (bulbs)	July	4 lb.	Autumn
White turnips (tops)	September	6 lb.	February–April

* In these cases the sowing is done under a corn crop and the feeding date applies to the following year. Italian rye grass may, however, be sown after harvest, for use in the next spring.

It has been said that catch crops are frequently sown on land that has had little preparation, and consequently that crop failure, partial or complete, is by no means uncommon; any statement as to the yield per acre to be expected must, therefore, be made with the greatest reserve. With this proviso, the following may be given as the very approximate yields of succulent fodder normally obtained per acre:

2 tons	3–5 tons	5–7 tons	8–12 tons
Barley	White turnip (tops)	Italian rye grass and trefoil	Rape
Trefoil	Rye	Oats and tares	White turnips (bulbs)
	Oats	Mustard	
	Crimson clover		
	Italian rye grass		
	Tares		

The food value per acre will also depend on the dry-matter content of the fodder consumed, and this will vary widely; in general, however, all the crops should be eaten off when in the young state.

It is, of course, possible to work out a system under which the normal root crops and seeds leys, in conjunction with catch crops, will provide a succession of succulent crops for feeding to livestock (cattle as well as sheep) throughout the whole year. The difficulty in making the system work in practice is that the weather is so very variable, and consequently crops will not grow according to plan; the fear of failure of some crops must drive the farmer to a generous provision of food, which means difficulty in getting it all consumed when he is fortunate enough to have every crop succeed. A complete system looks better on paper than it is apt to turn out in practice, and so it is that catch cropping tends to be restricted to a few odd fields, rather than to be the basis on which a farming system rests. Another difficulty that would attend a complete time table for the whole year would be the fact that these succulent crops rapidly become mature, and that then their feeding value declines at an alarming rate. From this it follows that only a small area should be sown at one time, especially if the crop is for feeding in summer, when it will be advancing rapidly in maturity; in summer a fortnight is as long as one crop, sown on one date, should be asked to provide keep. But the limitations of catch crops in providing a continuous supply of food must not be allowed to detract from their importance in coming to the rescue when food is particularly required.

CHAPTER II

MANURING

Of the elements which enter into the composition of plants nitrogen, phosphorus and potassium are most likely to be deficient, and the chemical fertilisers on the market supply one or more of these three, and no others. On some soils other elements may be lacking and this may give rise to diseases of crops. Much knowledge of these deficiencies has been gained in recent years, but the subject is somewhat specialised and the soils where pathological conditions arise from this cause are limited in area. The subject is interesting and, for the advisory officer, very important, but it is not susceptible of brief treatment; indeed, a little knowledge of it is definitely dangerous, as evidenced by the many wrong diagnoses made by those who have not the necessary specialised knowledge. These 'rare elements' will therefore be left out of this discussion. Liming will also be omitted although it ranks with drainage as fundamental to the maintenance of soil fertility. If land is acid or wet it is futile to spend good money on fertilisers to apply to it; such conditions must be corrected if farming is to be successful, but their consideration lies rather within soil science than crop husbandry. At the present time the farmers of England and Wales are nearing the figure of 7 million tons of chalk which are reckoned to be necessary as annual application to make good leaching losses and to correct within measurable time the remaining acid land. During and just after the Second World War much good drainage work was done but the pace slackens and if there should be an unhappy recession there will soon be much land incapable of giving response to manures.

The soil chemist can classify soils fairly accurately into four or five grades according to any one of the plant nutrients nitrogen, phosphate and potash and, if he is thoroughly familiar with the locality and the particular soil concerned, he can give very useful advice; he can at least tell which is the nutrient specially required. Soil analysis can be particularly valuable to the farmer who has just taken over land and, to established farmers, for periodical checks of nutrient content. Some quarter of a million soil samples are analysed each year by the National Agricultural Advisory Service but clearly it is impossible for every field of every farm to be tested annually. Soil analysis will detect serious deficiency in any plant food and to that extent it relates well to the results of field experiments (e.g. if a field is classified as 'very low' in potash, experiment will nearly always show that that field responds generously to potassic manuring), but it cannot tell the farmer what is the optimum economic dressing of fertiliser. For that he has to rely on accumulated experience and the large number of field experiments on the subject which have been reported. Finality has not yet been reached, but it has been shown that requirements of soils of different types vary much, that there are wide differences between soils of the same type, that individual crops have their own peculiarities and that the weather plays a large part in determining the increment of yield obtained from fertilisers. General rules on manuring are, therefore, of limited value, and a farmer has to consider all the circumstances germane to each case in his attempt to decide rightly which manures, and in what quantity, it will pay him to apply to a field.

There are three general principles in the science of manuring, which are all fairly obvious, but the implications of which should be thoroughly appreciated. They are:

(*a*) *The principle of limiting factors*. This is the same idea as that which is conveyed by the saying that the strength of a chain is that of its weakest link; applied to manuring, this means that the yield of a crop is limited by the plant nutrient that is in lowest supply, and that the shortness of one nutrient cannot be compensated by an excess of another. If, say, the land has a low content of phosphate it is no good applying nitrogenous or potassic fertilisers; phosphate in that case is limiting the yield, and without phosphate a full crop cannot be produced. The application of this principle would be simplicity itself if only it were possible to say which is the limiting factor, but it is just there that the difficulty lies. Results of chemical analysis will not usually be available and will not have sufficient precision, so the farmer has to guess the required nutrient, and in this he must call to his aid a variety of considerations. He must realise that particular crops usually respond best to particular nutrients; cereals and green crops usually respond best to nitrogen, leguminous crops, turnips and swedes to phosphate, while potatoes have a high demand for potash. He must appreciate the fact that heavy land is very liable to be deficient in phosphate and light land in potash, and he must also be guided by his experience of manuring his farm in past years; the condition of the field as affected by the previous crop, and the manuring that crop received, must duly be taken into account. These are the lines of thought that lead to the correct guess as to which nutrient is likely to be the limiting factor, but it may be, of course, that the factor which ultimately limits the yield of a crop is not a nutrient at all, but rainfall, sunshine, or something else equally outside the farmer's control.

(*b*) *The law of diminishing returns.* This is a simple corollary arising from the above, because as one limiting factor is removed the yield will be raised until another factor becomes the limiting one; it is only to be expected, therefore, that as a manurial dressing is increased each succeeding amount of manure will produce, in general, a diminished response. The point can be illustrated by the yields of wheat obtained on Broadbalk Field at Rothamsted; the following were the yields of some plots over a 74-year period:

	Yield of grain (bush./acre)
Unmanured	12
Minerals alone	14
412 lb. sulphate of ammonia	19
Minerals plus 206 lb. sulphate of ammonia	22
Minerals plus 412 lb. sulphate of ammonia	30
Minerals plus 618 lb. sulphate of ammonia	35

It is clear that the soil was short of all the nutrients, but nitrogen appears to have been the chief delinquent; a dressing of mineral manures only raised the yield by 2 bushels, whereas 412 lb. of sulphate of ammonia raised it by 7 bushels. It appears that nitrogen deficiency limited the yield to 14 bushels and mineral deficiency to 19 bushels, and it is seen that when both were applied the yield rose to 30 bushels, an increase considerably greater than the sum of the increases for each separately.

Further nitrogenous manure could only give an increase of 5 bushels, since apparently other limiting factors then came into play. The law of diminishing returns is illustrated by the increments obtained by increasing doses of 206 lb. of sulphate of ammonia where minerals were also applied: 8 bushels for the first, 8 for the second and 5 for the third. It must be noted, however, that these dressings of sulphate of ammonia were large ones, and even then the third 206 lb. of sulphate of ammonia produced an economic return; it is surprising, therefore, that figures from this field have been used to justify low farming, on the basis of the law of diminishing returns. Now that chemical manures are relatively cheap and plentiful, farmers should keep this law in the background of their minds, because there is no doubt that the great majority of them could with advantage increase the dressings they use, without suffering from greatly diminished returns; cheap chemical manure is one of the few trump cards in the farmer's hand—not to play it is to invite the usual result of 'bottling up the trumps'.

(*c*) *The danger of too much.* In some cases there is a danger that an excessive dressing may not only be wasteful, but may have a definitely harmful effect on the crop. This danger rarely, if ever, arises in the case of phosphatic or potassic fertilisers (except when they are drilled with the seed). It is the nitrogenous manures with which care is needed. Excessive nitrogenous dressings lead to lodging in cereals, favour disease in all crops, and lower the quality of root crops and of barley. In general it may be said that it is the nitrogenous manures which produce the greatest increases in yield, and therefore farmers are well advised who use them to the utmost; but it is unfortunate that they are also the fertilisers with which undue generosity of dressing is attended by serious risks.

From this brief consideration of the main principles on which manuring must rest, the next step should be to proceed with a discussion of the manures available to the farmer, but it is thought that this would lead to matters more suitable to a book on chemistry. It is proposed, therefore, to deal only with farmyard manure, which still occupies a place much higher than that of any other manure in British agriculture, and then only to consider the purely practical aspects.

FARMYARD MANURE

Farmyard manure has for centuries been the main fertiliser used on British farms and it might be supposed that by now its effects would be well known. This is not so, and, in fact, controversy on the subject, always keen, is particularly vigorous at the present time. There is a very vocal school of thought which holds that the greatest care should be exercised to return all possible vegetable waste to the land, having first converted it to humus by mixing it with animal residues and subjecting it to controlled decomposition in compost heaps. This school maintains that where soil is adequately manured with compost, and receives no 'poison chemicals', the plants that grow in it acquire a mysterious 'quality' which renders them immune to diseases and pests and that men and animals consuming such plants acquire a like immunity. Many observations are quoted in support of these extravagant claims but there is a significant lack of critical experimental data to uphold them. On the other hand, there are good farmers who think that farmyard manure or other sources of humus are not essential to the maintenance of fertility, that the high cost of handling these manures cannot be justified and that full crops can be grown year after year on chemical fertilisers alone. Many long-term experiments will be necessary to settle this dispute. Meanwhile, the great majority of farmers and agricultural scientists, whilst appreciating the value of chemical fertilisers and their power to produce good crops for 5 or even 10 years, are of opinion that organic manures (including as a very important case the ploughed-in turf) are essential if fertility is to be maintained indefinitely.

Among the valuable properties of farmyard manure, its physical effect is rated high, as being beneficial on all types of soil. On heavy land it facilitates cultivation, tests having shown that it appreciably reduces the draw-bar pull when an implement is dragged through the soil; further, when farmyard manure is incorporated into the soil it helps in crumb formation, and thus in producing tilth. On light land it tends to bind the soil and to prevent the latter from falling down to a powder, whilst its capacity to hold water is of absolute paramount importance on such soils. It is as a source of humus that farmyard manure has these extremely beneficial effects, and the effects are lasting; this is because of the straw which it contains, other organic materials, although providing humus, lacking the persistency of straw in the soil. It must not be thought that incorporation of organic matter, as in the occasional application of farmyard manure, will immediately and markedly improve the condition of a soil. In good farming the regular rotational dressings are cumulative but the physical effects of any one dressing are slight. The little difference may, however, be extremely important as when the small increase in water held by the soil just enables a young plant to survive a droughty period. Without it the plant might die, with it the plant may live to flourish when the rain comes.

It cannot be denied that farmyard manure has special valuable properties, but these advantages should not blind the reader to the fact that it is an important source of plant food. Since it has become possible to buy the latter in a sack the farmer has developed a tendency to forget this point, but in farmyard manure farmers are handling very valuable material, which merits careful attention quite apart from its mechanical and special properties. It is regrettable that the desirable care is not usually forthcoming.

In other countries dung heaps are frequently built on concrete, in which a grating leads to a tank to collect the liquid that drains away; the tank is fitted with a pump and the liquid manure, which is rich in plant food, is periodically pumped into a liquid manure distributor and applied to the land. Although some farms are similarly equipped in this country, the pump in most cases is allowed to fall into disrepair, the common opinion being that the value of the liquid manure is insufficient to justify the cost of application. It is difficult to agree with this view. Of course a dung heap in a wet district will yield a liquid that is much diluted with rain, and it is very rare for any sort of roof to be provided; apparently the extra cost of roofing is not considered worth while, and that may perhaps be the chief reason why an advocate for the saving and utilisation of liquid manure in this country finds himself as a voice crying in the wilderness.

Farmyard manure is far from being a standardised product, its value being affected by the four following factors:

(*a*) *The type of animal that makes it.* Cows and young growing livestock utilise nitrogen and phosphate from their food, either in the milk they produce or in their increase in body weight, whilst fattening animals retain hardly any; it follows that the dung of the latter is richer in plant food.

(*b*) *The ration fed to the animal that makes it.* There is an old belief, deeply ingrained in British agriculture, in the efficacy of manuring land 'through the animal'. This is sometimes held to justify the feeding of uneconomic rations to livestock for the sake of the rich dung they will provide; losses on the fattening of bullocks have sometimes been borne more or less cheerfully in the faith that they will be recouped in the crops for which the dung is used. Numerous experiments have been conducted to compare cake-fed and store-fed dung, and they have shown how uneconomic it is to feed an animal more than necessary, for the purpose of obtaining rich dung; some experiments have succeeded in demonstrating a slightly higher content of available plant food in cake-fed than in store-fed dung, and a consequent increase in yield in the first crop after its application, but the increase has always been small and in many cases has been impossible to detect. Whilst, therefore, it is worth while noting that the amount of plant nutrients in farmyard manure is affected by the food fed to the animal, it is much more important to realise the wastefulness of feeding the land through the animal. There are tables of analyses of feeding stuffs giving the manurial value of each food, that is, the value of the plant nutrients it will provide in the dung; this value varies roughly from 5 to 30 per cent of the cost price of the food, and it is surely bad policy to waste the other 70–95 per cent of the value for the sake of this small proportion. The point has become even clearer now that chemical manures are relatively cheap, and it is yet more indisputable where due care in handling the dung is not exercised.

(*c*) *How it is made.* Methods of making farmyard manure are very variable and considerably affect its value. Of outstanding importance in this connection is the amount of straw used. In some arable districts it is the practice to be very generous with straw, and sometimes livestock are kept merely to tread down the straw into dung; the yards in which such animals are kept are liberally strawed every day and, though no objection can be raised to this method, it must be realised that the farmyard manure produced is relatively poor in plant nutrients. Where straw is scarce, or has to be bought, it is used with great circumspection, and then the farmyard manure consists mainly of the actual dung. This point is linked to that of open or covered yards; in the former case more straw is needed and more plant food is leached from the farmyard manure whilst it is being made. Where the manure is made in yards or boxes the straw is put on top of the manure already made, and the mass built up during the winter to a depth of 3 or 4 ft.; in that case the lower layers absorb the liquid which drains through the surface layers and, if under cover, little will be lost. In milking sheds, stables and pig houses the manure is cleared out frequently and collected in a heap; in this case much liquid manure is lost down drains, whilst fermentation proceeds at a rapid rate in the heap.

(*d*) *How it is handled.* With the increase in wages during the past few years farmers have not unnaturally become averse to handling material twice, when once can be made to suffice; this is common sense, but some have harped on the idea to such an extent that it has become an obsession, and the question as to whether the second handling may not be worth while is then not even considered dispassionately. There are two common methods of handling farmyard manure made in yards or boxes. One is to allow it to lie where it is made all through the summer, and to cart it direct on to the field where it is required after harvest, leaving it there in small heaps ready for spreading, or using mechanical manure spreaders for the carting. The other is to cart it out from the yard to a large heap in the corner of the field in which it is to be applied, when convenient in late winter or spring, and to spread from this heap shortly before it will be ploughed in. It is the former method which is becoming more popular, because it only entails carting the manure once; it may also be urged in its favour that the method avoids the loss of the dark and offensive, but valuable, liquid which runs out from the large field heap. But there are certain drawbacks to the method. In the first place the carting must necessarily take place in most cases directly after harvest, because the yards must be cleared for the livestock which will be coming in for the winter some time in October. Farmyard manure is most plentifully used for root crops which are sown in spring, and for these crops ploughing cannot usually be done until about December, when the autumn rush of work is over; it follows that the method frequently leads to the manure lying in small field heaps for two months or more in the autumn. During this time there will be some loss of ammonia, and, though the plant food that is leached will be washed into the soil, it will only be washed into the small proportion of soil directly beneath the heaps, leaving very poor material for spreading later. It is very common to see, in the crop following such treatment of manure, rank growth clearly indicating the position where each little heap stood for so long; such uneven distribution of plant food over the surface cannot be justified, and no farmer would tolerate similar inequality of distribution of a chemical fertiliser. If the carting out, weeks ahead of ploughing, is done with manure spreaders so that the material is distributed all over the surface of the field, plant nutrients are leached into the soil more evenly but the loss of nitrogen as ammonia will be greater, particularly in dry weather. It is true that the method saves one carting, but it necessitates that being done about September when, on the ordinary mixed farm, there is much work to do; by the other method September carting is only necessary when the manure is to be applied for an autumn-sown crop, and even then the work is lighter, as it is much easier to fill from the heap than from a yard in which it has been thoroughly trodden, and is very tightly packed. But the most important difference between the methods is in the type of manure produced; by direct carting long dung is applied, whilst by carting first to a large heap the mass is aerated, fermentation set up and short dung produced. This preliminary fermentation reduces the straw to humus, and in its absence the same changes occur after the manure is incorporated with the soil; the latter may not have any important harmful effects where the manure consists mainly of dung and contains little straw, but in the reverse case the bacteria which decompose it will not find sufficient nitrogen for their purpose in the manure, and will take it from the soil. So the first effect of applying long dung may be to denude the soil of available nitrogen, and though the nitrogen is not actually lost it is locked up and will not be available again until the bacteria themselves decompose; it will be realised that this is the opposite of what is required, the farmer's aim being to render as much of the plant food available as possible. It is, of course, true that the fermentation in the heap will reduce the amount of the material, but the reduction need only be what will occur in the soil, within a short time of ploughing in, when long dung is applied. The best practice should ensure a controlled preliminary fermentation, so that desirable changes may take place and yet undue wastage be avoided. To cart manure to a large field heap and leave it there in a loose condition for many months invites excessive fermentation, and then the heap appears to fade almost entirely away. The forking of the manure from a yard to a cart will ensure sufficient aeration, and waste is avoided if care is taken to make the heap solid; the usual method of doing this is to build a heap of the 'cart-over' (or 'pull-over') type. A cart-over dung heap is usually rectangular, with a ramp at each of its short ends; the method of building is to lead a full cart up one of the ramps (one or possibly two trace horses being required), to empty the load at a convenient spot, and to lead the horse and cart down the other ramp. Alternatively, the heap may be square or round with one ramp leading up to it; then the full load is taken up the ramp and the cart travels round the top before returning down the same ramp. Unhappily the job of making a cart-over heap is not so easy with tractor and trailer as it is with horses and carts since wheeled tractors get stuck on the ramp as soon as it acquires a sharp incline. Tracklaying tractors can manage the job but few farmers have them and they are too expensive to waste on this intermittent type of work. In either type of cart-over heap the material is firmly packed down as building proceeds, and so the amount of air in the heap when it is completed is greatly reduced; thus fermentation is set up by admitting air on moving the manure, but will soon be checked in the heap through lack of air, and experience shows that something like the optimum amount of fermentation occurs. It must be admitted that the control of fermentation is by no means exact, but it is the best that can be reasonably expected under practical conditions. Against any undue loss that may occur may be set the important fact that the heat generated in the heap will be sufficient to kill weed seeds; this point must not be underrated, because many instances are seen of fields rendered foul with annual weeds by the application of fresh, long farmyard manure.

From what has been said above it will be gathered that the ploughing in of fresh straw is bad practice and considerably impoverishes the soil of nitrogen. On mixed farms such a procedure is not likely to occur, because straw is a valuable by-product of corn crops, and one much needed for the livestock; but there are some holdings on which livestock play a very small part, and then the question arises as to whether straw can be converted into humus without the animal to act as intermediary.

Those who have convinced themselves that humus in the soil bestows some mysterious benefit to plants advocate the rotting down of straw and all other organic wastes in compost heaps. These are carefully built in some form of trench, earth and animal residues, such as blood, dung and urine, being incorporated in layers, and the heaps are turned one or more times at intervals of 2 or 3 months; under this treatment the heap is converted into humus of undoubted value as a fertiliser. Arising from scientific work carried out at Rothamsted a process has been patented for rotting down straw, the method being to add ¾ cwt. of sulphate of ammonia to every ton of straw, to inoculate the heap with the necessary bacteria and to keep it wet; in a few months the straw comes to resemble short dung not only in appearance but also in manurial effect as measured by field experiments. More rough-and-ready methods were often adopted, with some success, with accumulations of straw during the War of 1939–45. Stacks were built with alternate layers, 1–2 ft. in depth, sprinkled with lime and sulphate of ammonia, and then the whole was thoroughly drenched with water; this last was possible when the country had a whole-time Fire Service anxious to test their appliances, but otherwise it would present almost insuperable difficulties. All these methods involve much labour. Where straw is converted into farmyard manure there is also considerable handling but there is, at least, the saving grace that the livestock which utilise the straw derive some benefit from it. It is doubtful whether composting will ever have an important place in British farming. Where superabundant supplies of straw are available, cattle-yards are strawed very generously so that the livestock may tread down the maximum amount; if a light sprinkling of sulphate of ammonia is given to the yard after each strawing decomposition is quickened and shorter dung is produced.

It has been stated that the content of plant food of farmyard manure is very variable; the reader must, therefore, regard any figures for its composition as no more than rough guides. Analyses show that the amount of nitrogen (N_2) contained in it may vary from 0·4 to 0·8 per cent, the phosphate (P_2O_5) from 0·2 to 0·5 per cent, and the potash (K_2O) from 0·4 to 0·8 per cent. Ten tons of short dung contain about as much nitrogen as 5 cwt. of sulphate of ammonia, as much phosphate as 3 cwt. of superphosphate, and as much potash as 2½ cwt. of muriate of potash. Its value may therefore be calculated as follows:

	£	s.	d.
5 cwt. of sulphate of ammonia at £14 per ton	3	10	0
3 cwt. of superphosphate at £8 per ton	1	4	0
2½ cwt. of muriate of potash at £17 per ton	2	2	6
10 tons of farmyard manure	£6	16	6

whence the value per ton is calculated as roughly 13*s.* 6*d.* Something should be subtracted from this figure because of the losses that occur before the manure is finally ploughed into the land; it is impossible to suggest a figure for this, and in any case argument on the question is unprofitable, because of the many uncertainties in appraising the value of the farmyard manure. One factor which weighs heavily against it is the cost of carting and spreading it, this cost for an ordinary dressing being generally in excess of £5 per acre, as compared to a few shillings in the case of a mixture of artificial fertilisers. On the other side must be set the valuable physical properties mentioned earlier, and also the fact that part of the plant food in farmyard manure is readily available and part becomes available in course of time. The provision of available food from a dressing may be very long continued. A high proportion of the plant food in a dressing of farmyard manure is undoubtedly used up in the first year or two after its application, but there is something left over, which, together with the mechanical effects, benefits the land for years to come; it is also true that farmyard manure contains a wide variety of elements, and so its periodical application prevents the total exhaustion of the soil of any of the elements which are known to occur in plants, but which are normally assumed to be present in the soil in sufficient quantity. It is not surprising, therefore, that the general experience is that yields of crops tend not only to be higher, but also to be more uniform from year to year, where farmyard manure is applied regularly, than where it is not. The advent of chemical fertilisers has, of course, been a very important forward step in agriculture, but farmyard manure remains, and probably will remain, the basis of a sound manurial policy.

DRESSINGS OF FARMYARD MANURE

The fundamental importance of farmyard manure makes it highly desirable that all arable fields should receive roughly their fair share of it, and the argument has been advanced earlier that one of the benefits of a rotation is that it ensures that the crops for which dung is applied shall be taken on each field in due course. By any system of cost accounting a dressing of farmyard manure is very expensive; the cost of application alone is very high. When this is appreciated, together with the fact that a large proportion of the return from it is obtained from the first crop after application, it will be realised that common sense requires that dung shall be applied not only for a crop that will respond to it, but also for one whose value per acre is high enough to pay for it; in general, practice is in conformity with this point.

The following crops should always receive farmyard manure, if there is any available:

Potatoes,
Sugar beet,
Mangolds,
Beans,
Spring cabbage and other vegetables,
Soft fruits.

Potatoes, sugar beet and cabbage are high-value cash crops which respond well to dressings of dung, whilst the mangold crop is capable, when generously treated, of producing huge yields of stock food. On heavy land root crops play a very small part, and beans is the crop which normally receives farmyard manure; beans grow well with plenty of humus in the soil, and it is a sound principle to treat them generously, both for their own sake and also because a successful bean crop is an excellent preparation for wheat.

The following crops come next in priority of claim to dung:

Kale,
Turnips and swedes,
Wheat.

Kale undoubtedly responds well to farmyard manure, and is usually accorded a dressing when there is any to spare, which, however, is often not the case. The position with turnips and swedes is that the yield is highly dependent on rainfall, and consequently better crops are obtained in wet than in dry districts. In wet districts farmyard manure is normally applied, but in dry districts the danger is that, despite a dressing of dung, the crop may be a poor one, and much of the value of the dressing will remain in the soil, which will then produce poor-quality barley, that being the crop which, in dry districts, normally follows roots. Consequently in East Anglia turnips and swedes rarely receive farmyard manure, whilst in Scotland they usually do; with kale the same is true but to a less extent, as kale is rather deeper rooted than turnips and swedes, and hence not so utterly dependent on rainfall. The position of wheat in this list is rather strange, because, though it is a crop which requires land in fairly good condition, it would not be likely to repay the expense of a dressing of farmyard manure, except in the infrequent case where it is grown as a second corn crop. Nevertheless, wheat grown on light land in the eastern counties often receives a dressing of dung; in particular, it is the wheat following seeds which receives the dung. This is to be traced to the old Norfolk four-course rotation. It has just been said that the root break of that system would not usually receive dung, and neither would the barley nor the seeds; there remained, therefore, only the wheat, and the almost universal practice was to 'muck the ollands', that is, the old leys just before they were ploughed up for wheat. On poor light land there is little danger of the wheat lodging, and the practice has the advantage of ease of carting; further, there is evidence that when a ley is ploughed up the turf will rot down better if farmyard manure, which contains some readily available nitrogen, be incorporated into the soil with it. The old practice is, in fact, in tune with the most modern scientific views, but now that so many farmers have modified the Norfolk four-course system, the wisdom of this practice may be doubted; where, for instance, sugar beet is included in the rotation that crop will usually respond more generously to farmyard manure than will the wheat.

The following crops rarely receive farmyard manure:

Rape,
Mustard,
Peas,
Oats,
Barley,
Leys.

The first two crops in this list undoubtedly respond well to farmyard manure, but they are usually grown cheaply and are regarded as catch crops on which little money is spent; mustard, indeed, is largely grown to plough in as a substitute for farmyard manure. Peas, also, undoubtedly respond to dung, but on all except the poorest land a dressing tends to produce too much haulm, with earlier lodging, few pods and a more vigorous growth of weeds. Oats is not a high-value crop, but as it is frequently grown after another corn crop it does sometimes receive a dressing of dung; this, however, is only when the needs of other crops have been satisfied. Barley grown with any hope of selling for malting should never receive dung, as the latter cannot be spread evenly on the ground; despite the greatest care there will be fair-sized lumps to plough in, and these will produce unevenness in the crop, and will prevent the uniformity which is so essential in a high-quality sample. Furthermore, a dressing of farmyard manure is equivalent to a fairly generous dressing of chemical nitrogenous manure, and consequently tends to produce steely barley with a high nitrogen content. It is not very uncommon to apply farmyard manure to a short ley in the autumn after seeding, but the increased growth is more apparent than real, that is, self-sown cereals are encouraged to grow luxuriantly but there is little effect on clover; the practice is to be deprecated, if any of the crops mentioned earlier are to be grown on the farm.

Permanent grass not infrequently receives dung, but that is on farms in the west of the country which include no arable land. In such circumstances application to grassland is inevitable and a considerable increase in growth is produced, particularly early spring growth, in which connection the shelter provided by the dung may be important; application during the winter to fields intended for hay in the following summer is advisable if farmyard manure is to be applied to grassland. If a field is cut for hay year after year—a piece of really bad farming which cannot sometimes be avoided—farmyard manure may reasonably be given to make up for the constant drain suffered by the field; but grassland is normally grazed for much of the time and it thereby receives large amounts of urine and dung. Farmyard manure will produce from arable crops increases of yield much more valuable than the increase obtained from grassland, so that in mixed farming grass rarely receives it; occasionally, however, where the straw in the farmyard manure was from a very foul field, it may be good to apply it to grassland, because there the seeds of arable weeds will not establish plants.

Apart from the rare cases where a few odd tons of farmyard manure are available or left over, and are spread more with the idea of getting rid of them than of manuring the field, the minimum dressing applied is about 8 tons; less than this is impossible to spread, and in general the principle is to do one or more fields well, rather than to spread the manure evenly over all the land which is felt to require it. Where more than twenty tons is given, generally only in market gardening, it is commonly applied in two doses, one in the autumn and the other in the following spring. A normal dressing is about 10–12 tons per acre, though green crops, mangolds and potatoes may reasonably be allowed more, if there is plenty available.

In the days when horses were the source of power on British farms the method of carting farmyard manure on all save the smallest farms was to organise a 'set'. Two to four men loaded continuously from the heap or yard and one man emptied the carts in the field (a relatively light job as it only involved pulling the manure off the back of the tipped cart with a muck-rake); according to the distance the manure was being carted three or more horses and carts were required with one or more boys to lead the horses. At any one moment one cart would be loading, one emptying and the others in transit giving a steady continuity to the whole process. Commonly the number of heaps was constant at 100 per acre, variations in heaviness of dressing being obtained by changing the number of heaps per load; a load was from 15 cwt. to 1 ton and the number of heaps per load from five to ten. The heaps were usually put with 6 yd. between them in rows 8 yd. apart; subsequent spreading was easier when the heaps were 'on the square' but some carting was saved by having the inter-row slightly greater than the intra-row distance. Spreading the manure from the heaps was a slow job and it was reckoned as half a day's work for a man to spread the heaps on an acre; this was often done by men at odd times and most farmers were rightly particular that it should be done evenly and the manure was not left in lumps.

At the present time almost every conceivable organisation of manure-carting can be found in the country. The mechanically minded farmer may point the finger of scorn at the slow pace of the horse but the organised set did provide a continuous process with everyone working all the time. If the horses are displaced by one or two tractors (relatively few farms have more) the whole thing degenerates to 'lob cart' with strong men spending a large part of their day riding to and fro between the heap and the field. Tractors and trailers have a much greater tendency to get stuck in the mud than horses and carts and since trailers cannot in general be tipped it takes nearly as much labour to unload as to load. Mechanisation has greatly increased the efficiency of most operations, but its effect on carting farmyard manure has been disastrous. This is the more galling because there are machines for performing each of the separate operations involved—loaders which are quick and save hard hand labour, spreaders which distribute the manure in a very short time and much more evenly than can be achieved by hand and distributors that deal effectively with heaps already spaced in the field. But no farm of moderate size can carry the whole gamut. The problem may be solved by the use of machines that will not be limited to one job, as, for instance, elevators that can handle sacks and bales as well as farmyard manure. Contractors may equip themselves to undertake the whole thing but modern loaders work so quickly that five or six tractors and spreaders may be needed to keep them working efficiently. Meanwhile, the position on most farms is extremely unsatisfactory, with more man hours per ton required for the job than were required a generation ago.

For autumn-sown crops farmyard manure is carted out immediately after harvest and is spread and ploughed in as soon as possible. For root crops it is also generally carted out after harvest, and is then often left in the small field heaps until the autumn rush of work is over, when it is spread and ploughed in. It has already been said that this long wait on the field is undesirable, and, in fact, experiments have definitely shown that appreciable losses occur, and that heavier yields follow immediate spreading and ploughing. Where the manure has been carted first from the yards to a large field heap the second carting may be deferred until just before ploughing, but, of course, the carting will usually be heavier, as the land may be expected to become softer as Christmas approaches; in that case it is worth while waiting for a severe frost, so that the carting may be over frozen ground. It is better to apply farmyard manure for root crops in autumn than in spring, because in the latter event the land will have to be ploughed in spring to cover it; the general principle in preparing a seedbed for roots is to plough in late autumn so that the land may receive a full winter weathering, and it is very undesirable on all but the lightest soils to bury the winter mould by ploughing in spring. In the case of sugar beet there is the added consideration that the application of dung in spring will tend to make the beet fangy. Where a crop is being sown in ridges it is not uncommon to open out the furrows, apply the farmyard manure, and split the ridges over it. This is still a common practice of potato growers, the argument advanced for it being that the seed tubers are planted directly on the manure, which, in addition to supplying plant food, will hold up moisture for the plants. But there is a number of arguments which can be adduced against the practice. In the first place, the potato roots are not restricted to a small space just under the sett: secondly, unless the dung be in the short, well-rotted state it will tend to make the ridge more open, and may thus favour the loss of moisture in a dry season: thirdly heavy carting will fall on the furrows where the manure is to be spread, that is, under the position of the seed tubers, and consolidation at that point is the reverse of what is wanted: lastly, much labour is required when potatoes are planted, and the demand is appreciably increased when manure has to be carted and spread then, instead of during the slack period in the autumn. The balance of argument is probably against the practice of applying dung in the ridge, and that procedure is less common than formerly. Mechanical potato planters, which open the ridge, drop the sett and ridge up all at one passage, are becoming more and more common, so that application of farmyard manure in the ridge must necessarily decline still further. Where it is done, the method is to put heaps rather smaller than normal ones in the furrows, along which they are spread. On grassland, farmyard manure may be applied at any time, but it is usually done during the slack winter months, the wetter periods being avoided so that the field may not be cut up; after spreading, the field may be harrowed lightly to disintegrate the lumps further, and it is surprising how quickly the manure disappears, through the action of beating rain and of worms.

CHEMICAL FERTILISERS

At the present time it is very difficult to decide on the notation to use in discussing chemical fertilisers. A full account of the different fertilisers lies outside the scope of this book, but it will be realised that there are several for each of the three major plant nutrients. Those which contain only one of the nutrients are referred to as 'straight' fertilisers. These, within each class (e.g. the nitrogenous ones) are interchangeable up to a point and the farmer is naturally attracted by the one that costs least per unit of the nutrient. But there are factors other than price to consider. For example, on land with a low lime reserve nitro-chalk and basic slag should be used rather than sulphate of ammonia and superphosphate. It has been the general practice to speak of complete dressings in terms of sulphate of ammonia for the nitrogen, superphosphate for the phosphate and muriate of potash for the potash and to suggest the appropriate amount of each of these for a particular case; a farmer wishing to apply all the nutrients at once would mix the straight fertilisers together according to the prescription.

But practice has changed in recent years and now some two-thirds of all the chemical fertilisers purchased by farmers are compounds, containing two or all three of the major nutrients, and few farmers make up their own mixtures. Whether this is a desirable change or not will be considered later but that is the present position and the tendency is for the importance of compounds to become even greater. In these circumstances discussion of appropriate dressings in terms of straight fertilisers would be rather divorced from practical realities. Most manure merchants offer a range of compounds so that the total number of different ones available to the farmer is large and there is wide variation both in concentration and in the ratios between nutrients. The law requires that the percentage composition of a fertiliser shall be declared and the percentage of nitrogen is given as such (N_2) whilst those of phosphorus and potassium are given as their oxides (P_2O_5 and K_2O respectively); it would seem more sensible to give the content of each as the element itself but that final step in rationalisation is unlikely to be taken in the near future. When, therefore, the composition of a compound is given as 6–9–12 this means that it contains 6 per cent of nitrogen, 9 per cent of phosphate and 12 per cent of potash, that always being the order in which the contents are stated. One hundredweight of this compound will contain 0·06 cwt. of nitrogen, 0·09 cwt. of phosphate and 0·12 cwt. of potash and the corresponding amounts for two or more cwt. are derived by simple multiplication.

In what follows this notation will be adopted, dressings being given as decimals (to two places) of hundredweight for each nutrient. The notation is admittedly inelegant but very little experience is necessary to master it and then it is simple and unequivocal; it is the notation used in published statistics and is becoming more popular in scientific papers. One example may help in clarification. Suppose it is desired to apply per acre 0·60 cwt. nitrogen, 0·40 cwt. phosphate and 0·50 cwt. potash to a particular crop and the farmer has in stock a fertiliser of 6–12–15 composition; 3 cwt. of this compound supplemented by 2 cwt. of sulphate of ammonia (21 per cent nitrogen) will be just about right. It is true that the phosphate will be 0·04 cwt. and the potash 0·05 cwt. too low, but recommendations can never be precise and such discrepancies are quite negligible. It is likely that in future fertiliser merchants will state not only the contents but also the plant food ratio, the latter to the nearest quarter. Thus the above compound would have a ratio of 1 : 2 : 2½ and these plant food ratios will help a farmer in selecting from his merchant's list the compound which will most nearly supply the nutrients required for a particular case in the right proportions.

NITROGENOUS FERTILISERS

The supply of nitrogenous plant food in the soil is maintained in a variety of ways. Farmyard manure is one important source, and in its absence or scarcity other organic manures are commonly used, whilst plant remains, particularly of leguminous crops, that are ploughed in contribute their quota; in addition there are in the soil beneficent free-living bacteria which fix nitrogen from the air, and some small amount of nitrogenous compounds is dissolved in rain. All these sources together provide valuable supplies, but nevertheless chemical nitrogenous fertilisers have proved themselves capable of giving large increases in crop yields. The four chief nitrogenous fertilisers are sulphate of ammonia, nitrate of soda, nitro-chalk and calcium cyanamide. Of these the first has a good condition (a word which, when applied to chemical fertilisers, connotes all the characteristics which make for ease in handling), and is not likely to be leached through the soil, but is not to be advised on a soil with a low lime reserve. Nitrate of soda is normally rather lumpy and sticky, though it is available in granular form, but has no tendency to accentuate acidity; it is liable to leaching, and is not to be recommended for frequent application to heavy soils, which it tends to make more sticky. Nitro-chalk is rapidly gaining in favour, because of its excellent condition and of the fact that it adds in a humble way to the lime reserve of the soil. As the nitrogenous compound it contains is nitrate of ammonia, half of its nitrogen is readily available, whilst the other half becomes available more slowly. Calcium cyanamide is not used very widely, largely because of the discomfort suffered in applying it, but it is sold in granular form to which no exception can be taken; in the powder form it may be valuable as a top dressing in that it will kill weeds such as charlock.

The following are the main general effects of nitrogenous fertilisers:

(*a*) *They stimulate the growth of the green parts—leaf and stem—of the plant, imparting a characteristic dark green appearance.* In some crops—kales, cabbage, etc.—this added vegetative growth is a direct return, because it is the green parts of the plant which are required; in other cases, where the valuable part of the plant is the seed or root, increased yields are also generally produced, because maximum production will not be attained without full vegetative development.

(*b*) *They increase liability to lodging in cereals.* The more generous growth of straw, which nitrogenous manures evoke, renders the crop much more liable to be lodged, and this fact limits the amount of nitrogenous manure which can profitably be applied to cereal crops. New varieties have, in general, better standing straw than the old ones and one of the plant-breeder's main contributions is that he has made possible a more generous use of nitrogenous manure.

(*c*) *They stimulate tillering in cereals.* This effect is very frequently observed and is regarded by farmers as definitely beneficial; it will be suggested later, however, that increased tillering does not necessarily lead to increased yield.

(*d*) *They delay maturity.* The stimulation of vegetative growth prolongs that phase of the plant's life, and consequently the ripening off process is deferred. This point is particularly important in the case of sugar beet, since sugar percentage rises as maturity is approached.

(*e*) *They make crops more susceptible to disease.* This is a very serious drawback to nitrogenous manures and applies to many crops; cases that may be mentioned are potato blight, mildew in turnips, swedes and cereals, rust in cereals. Presumably the effect is chiefly due to the fuller vegetative growth enabling fungi to spread more rapidly through the crop, but it is possible that the luscious type of growth produced may be more easily entered by the fungi.

(*f*) *They lower quality.* This is not universally, but generally, true. Excessive nitrogenous manuring of barley may increase the nitrogenous content of the grain, and that is tantamount to saying that the malting quality is lowered; on the other hand it is uneconomic to manure barley over-generously with nitrogen with the object of raising its feeding quality. The lower sugar content of sugar beet induced by heavy nitrogenous manuring may be ascribed to a delay in maturity, and it is probable that the same thing happens with mangolds, though figures to establish the point are not available. The general tendency is to lower the dry-matter content of roots, and this impairs their keeping quality. Nitrogenous manures are believed to tend to produce watery potatoes of low cooking value, but the proof of this is lacking. The lowering of quality of grassland is a very important instance, but that is chiefly due to a change in the botanical composition of the herbage.

Four of the above six points consist of warnings against the dangers attendant on the over-generous allowance of nitrogenous manures; it is, therefore, very important that there should be a full appreciation of the fact that these fertilisers are very potent in increasing yield. To dispense with them because of the dangers inherent in their use is a poor-spirited and quite unjustifiable policy, but to take the greatest care not to exceed the safe amount is imperative.

Very much experimental work has been devoted to the nitrogenous manuring of winter wheat, in which two separate questions arise—first, as to whether it is safe to apply any, and secondly, in the event of the reply to the first being in the affirmative, as to when it should be applied. In connection with the first it must be remembered that the place of wheat in most common rotations is a favoured one, in that it follows fallow, beans, peas, seeds, potatoes, or possibly roots, all of which leave the land in good heart. In all these cases the land should be rich in nitrogen, at least when the wheat is sown, and any nitrogenous dressing may, by increasing lodging, do more harm than good. Wheat should receive a top dressing of nitrogen in spring in all cases save where it is on land of very high fertility. The practice has been to apply about 0·20 cwt. and that was frequently given in late February or March. It will be seen below that better results follow application in late April or May and new varieties of wheat have made it possible to step up the dressing. The object in regard to nitrogenous manuring is to put on as much as possible without causing the crop to lodge. Cereals respond generously in yield but lodging is a serious matter as it increases the cost of harvesting. Modern combines are very adept at picking up lodged crops but cutting is necessarily slower; a more important point is that lodging often leads to poorly filled ears, particularly when it occurs some weeks before the crop is ripe. Cereal plant breeders have been very successful in recent years in producing better standing varieties than the older ones and hence farmers can be more generous with nitrogen. In most cases a dressing of 0·40 cwt. is quite safe, whilst some go as high as 0·75 cwt.

Nitrogenous dressings may increase the yield of wheat in two ways. Early application of nitrogen produces more tillers, that may carry ears at harvest. Very few, if any, tillers that are produced after the middle of March live to carry ears; if the maximum number of ears is to be evoked the dressing must, therefore, be early, and application to the seedbed, at the time of drilling, is the most effective. Application at the time previously favoured by farmers—late February or March—has little to recommend it; the nitrogen will then be largely wasted in producing abortive tillers in April, which will die down in May and June. If a dressing be given after the middle of April the plant will be passing out of the stage of growth when tillers are produced, so that wastage in that respect will be avoided; this late dressing of nitrogen may, however, benefit yield by increasing the weight of grain per ear, which it accomplishes by causing the setting of a few more grains at the top and at the bottom of the ear. May dressings are often very effective in this respect, so long as there is sufficient rainfall to wash them into the ground; dressings later than May are rarely effective in increasing grain yield, their only result being a greener plant with considerable delay in maturity. The position, then, may be summarised as follows: early dressings produce more ears, late dressings produce bigger ears, and intermediate dressings are liable to be wasted. The question then arises as to whether it is possible to get the best of two worlds—to give an early dressing to increase the number of ears, and a late dressing to make the ears bigger—and a number of experiments has been directed to this point. It may be said that, in general, the double effect cannot be obtained, the indication being that time is more important than amount of application; furthermore, it has been found that the relative efficiency of early and late dressings is utterly dependent on the weather of the season. If the winter proves wet, autumn dressings suffer from leaching and produce but little increase in yield; in that case late dressings are extremely beneficial. If the winter proves dry, the autumn dressings lead to a large increase in yield, whilst late dressings are ineffective, presumably because the early dose has not been washed out of the soil. The optimum results, therefore, are obtained by applying the nitrogen in autumn in a dry season and in late spring in a wet one, but this finding suffers in its utility from the fact that it is impossible to forecast the weather of the winter when the wheat is drilled in autumn. After a crop which leaves the land rich in nitrogen, or where farmyard manure is given, an autumn dressing of nitrogen is unnecessary and should be withheld; in other cases 0·20 cwt. is ample. Except on the richest soils a spring dressing will be amply repaid and 0·40 cwt. should be given in late April or early May, some farmers applying twice this suggested amount. After a very wet winter the crop may look miserably poor and yellow and with the strong strawed varieties 0·40 cwt. in March may be given with little risk of throwing the crop down; whether or not this is done the May dressing should be given. As regards which is the best manure to use, any dressing given before the end of March should be sulphate of ammonia or calcium cyanamide, because they are less likely to suffer from leaching; for later dressings the more readily available nitrate of soda or nitro-chalk should be used. In May the crop will, of course, be fairly high and the manure is then best applied broadcast, by men straddling the rows; for this case nitro-chalk is especially suitable because it is gritty, and the little lumps will bounce off the plants on to the ground.

Little experimental work on the nitrogenous dressing of oats has been conducted. This crop frequently follows another cereal and, except on the best soils, it will then respond well to nitrogenous fertilisers; the optimum amount of the dressing depends on the variety of oats, but even with the new stronger-strawed varieties the dressing should rarely exceed 0·40 cwt. per acre. It is to be presumed that in the case of winter oats the same considerations as to time of application arise as with wheat, but experimental confirmation of this is lacking at present. Spring oats normally receive their dressing on the seedbed, but it may be delayed until, or first decided upon by the appearance of a crop in April or even later.

The value and dangers of nitrogenous manuring are particularly well defined in the case of barley. This is a shallow-rooted crop which responds readily to nitrogen. Where barley follows another cereal the farmer cannot afford to neglect the application of a dressing. On the other hand, excessive dressing may ruin the malting value of a sample; even if it does not induce lodging with consequent badly filled and discoloured grain, it will tend to raise the nitrogen content of the grain and to make it flinty. It has been established that where a field is in low condition, nitrogenous manuring will, up to a point, give a generous yield increase without any of the above harmful results, but beyond that point serious lowering of quality occurs; the difficulty is to decide how much it is safe to apply. The general opinion is that where barley follows folded roots the land is rich enough, is, in fact, often too rich already, and that therefore no nitrogenous manure should be given; even this needs qualifying for a very wet winter, and when the roots are folded early, because then most of the nitrogenous residues from the sheep will have been washed out of the soil, and dressing may be wholly beneficial. After carted roots it will usually be safe to apply 0·40 cwt. of nitrogen per acre. If the barley follows corn some nitrogenous manure should be given, except on the richest soils, and experience indicates that 0·40 cwt. is the optimum dressing per acre, though more may be risked on poor, thin soil. As regards time of application there is little doubt that it should be the time of drilling; a top dressing later than April will certainly lower the malting value of the sample, experiments having shown that the later the dressing the more likely is this effect to appear. Where the barley is certainly going to be used for feed the only limit is the danger of lodging and the above dressings may be increased by as much as 50 per cent; it must be repeated that maximum yield is the aim, and not improvement in feeding value, as heavy nitrogenous manuring is a very expensive way of producing protein in barley.

Pulse crops should not, in general, receive nitrogenous fertilisers. It is true that leguminous plants benefit from them in the very early stages of growth, but as soon as their roots are established, and nodulation occurs, the bacteria provide an ample supply of nitrogen. Beans usually receive farmyard manure, which offers them an adequate amount of nitrogen to get them established.

Root crops respond well to nitrogenous manure, and normal dressings are rather higher in their case than with cereals. Sugar beet should receive, in addition to farmyard manure, 0·60 cwt. of nitrogen per acre; higher dressings than this are often given, and are undoubtedly repaid if the season proves wet, but great generosity increases the weight of tops rather than the weight of roots, and the sugar percentage of the latter is apt to be lowered. It is very common to give a small dressing (say 0·20 cwt. of nitrogen per acre) as nitrate of soda or nitro-chalk to sugar beet immediately after singling. The reason for this is that the crop always looks miserable just after the majority of plants have been chopped out, and those left are lying flat on the surface, and the farmer's natural instinct is to give something to help the crop pick up; it is surprising, however, how quickly the crop recovers from singling without any help, and experiments have failed to show any benefit from reserving part of the nitrogen for that time, as against applying it all on the seedbed. It has been shown quite definitely that nitrogen should not be applied later than immediately after singling, because subsequent dressings seriously lower sugar percentage. Mangolds normally receive about the same amount of nitrogenous manure as sugar beet; it may be that slightly greater generosity with mangolds is justifiable, because of the huge weights per acre which they are capable of producing. There is evidence that for sugar beet and mangolds the most efficient nitrogenous fertiliser is nitrate of soda, because these plants can use the soda as an alternative to potash; experiments have shown that sugar beet and mangolds respond less to potassic manure when the nitrogenous one is nitrate of soda rather than sulphate of ammonia. The point has been proved beyond doubt but the gain in efficiency is not enough to make up for a markedly higher unit price of nitrogen in nitrate of soda. Swedes and turnips should not receive heavy dressings of nitrogenous manure or they tend to suffer more from mildew and will not keep; if farmyard manure is applied, probably it is better to omit nitrogen, whilst in its absence the dose should be limited to 0·50 cwt. per acre. In wet districts rather more generous treatment with nitrogen is justifiable. If the soil is at all inclined to be acid, sulphate of ammonia should not be used, because of the danger of finger-and-toe disease.

The kales should be generously manured with nitrogen even if farmyard manure is given. The common practice is to apply from 0·20 to 0·60 cwt. per acre, but dressings up to 2·00 cwt. per acre have been given, and have justified themselves, in experiments. A good practice is to apply 0·60 cwt. as sulphate of ammonia on the seedbed and to give 0·40 cwt. as nitro-chalk as soon as the crop is satisfactorily established; some farmers will apply a further 0·40 cwt. a month later. It is probable that kohlrabi will repay equally generous treatment, since the bulb of that crop is a swollen stem, but actual experimental evidence on the point is scarce. Cabbages and related crops of the market-gardening type receive large dressings of nitrogenous fertiliser. Cabbage for spring should be transplanted in autumn on to land that has been well dunged, and should receive 0·80 to 1·00 cwt. per acre of nitrogen in the early spring; the difficulty about the latter is that if the manure is broadcast much will be caught by the leaves of the plant, which will suffer from dehydration at the points of contact, and this can only be overcome by distributing the manure round the plants individually, or along the inter-row spaces. Rape rarely receives nitrogenous manure, but undoubtedly responds well to dressings up to 0·60 cwt. of nitrogen per acre; mustard is usually grown very cheaply when it is intended as a green manure, but when it is grown for seed a dressing of up to 0·30 cwt. per acre of nitrogen (possibly in addition to farmyard manure) is sometimes given.

Potatoes respond well to nitrogen and dressings of 0·40 to 1·00 cwt. per acre are common; the high value of the crop will justify liberal treatment, but the tendency to increase susceptibility to disease, and to lower quality, puts a check on the amount that can be applied economically. There is a curious belief in some districts that early potatoes will give no appreciable return from chemical manures, but experiments in those districts have shown the belief to be entirely fallacious; in other districts it is the common practice to treat early potatoes at least as generously as main-crop potatoes.

Short leys generally have a leguminous plant as their chief, if not their only, constituent, and consequently receive no nitrogenous manure. In some cases Italian rye grass or timothy is grown alone, and then the production may be considerably increased by the application of 0·50 cwt. of nitrogen in early spring.

Long leys and permanent grass give large responses to nitrogenous manures, but there is grave danger of making the grasses grow rank and coarse, and of killing out the clover. There is undoubtedly opportunity for the utilisation of these manures on pasture, but their use should be accompanied by careful management. They can be applied to lengthen the growing period, up to 0·50 cwt. per acre being given in February or August; these dressings will have no harmful effects if they are followed by heavy grazing in April-May and autumn, respectively. To encourage a heavy hay crop by nitrogen usually has a disastrous effect on the botanical composition of the sward. It has been shown that the application of nitrogenous manure to a field about a fortnight before it is cut for hay increases the crude protein in the herbage and, with high protein foods expensive, farmers have been urged to adopt this practice. Crude protein is determined by finding the total nitrogen in the sample and multiplying by 6·25, but some of the nitrogenous compounds are not proteins and the real food value of these is unknown; most farmers have wisely hesitated to adopt the practice which requires further scientific work and practical experience before it can be recommended with safety.

The cost of purchasing, carting and applying 0·20 cwt. of nitrogen is in the neighbourhood of 17*s*. It would be satisfactory if it were possible to state the amount and value of the return to be expected from the various crops, but the impossibility of giving reliable figures will be appreciated; the state of the soil and the season are powerful factors influencing the increment obtained. Assuming that the farming is generally good, that other plant foods are available in the soil in sufficient amount, that disease and lodging do not seriously impair the crop, and that the nitrogenous dressings are of reasonable amount, the following are the sort of returns which may be expected, over a run of years, from 0·20 cwt. of nitrogen:

Cereals	3–8 bushels
Sugar beet	½–1 ton.
Mangolds	1–2 tons
Turnips	½–1 ton
Kale	1–3 tons
Potatoes	½–1½ tons

Even if the lower figure be taken, the increment is of sufficient value in every case (except possibly the turnips) amply to repay the cost of application. But much is involved in the above proviso that the nitrogenous dressing should be of reasonable amount, both in regard to the richness of the soil and to the crop.

PHOSPHATIC FERTILISERS

The main source of phosphatic plant food is the soil itself, but there is very great variation between soils in this respect. In some regions of the world, notably in parts of Australia and New Zealand, great deficiency in phosphates is found, and in some districts of this country phosphatic fertilisers are of first importance. It can be said that it is usually the heavier soils that are most likely to be deficient, but a more detailed classification than heavy and light is needed before sound advice can be given. A competent soil chemist with local knowledge can now say with some certainty whether or not a particular field is likely to respond to phosphatic fertiliser. It should be realised that farmyard manure—especially that produced by milking cows or growing stock—is not very rich in phosphates; it is, in fact, unbalanced, and complete reliance on dung can only lead to the gradual lowering of the phosphate status of the farm.

The chief phosphatic fertiliser is superphosphate, for which it is claimed that the phosphate is in a readily available form; this is no doubt true, but as soon as the manure is incorporated in the soil chemical changes take place with results that are very imperfectly understood. The older type of superphosphate was frequently in poor condition and was inclined to be sticky and even to contain free acid, but in recent years a granular form has been placed on the market, and this is in excellent condition. Mineral phosphate is in good condition, though dusty, and is very cheap per unit of phosphate, but its very low solubility makes it worthless in many situations; where the rainfall is heavy and where the soil is inclined to be acid it is valuable, and then it provides a very cheap source of phosphate. Basic slag also has its phosphate in a form only very slightly soluble, and both this manure and mineral phosphate are sold in a very finely ground condition to increase the surface area of the particles; basic slag is used principally on grassland, but considerable amounts are applied to arable land, especially where the soil has a low lime reserve. Basic slag contains variable, but considerable, amounts of the so-called minor elements and special virtue may be claimed for it, not only for its lime content, but also as a supplier of these and hence as tending to prevent deficiency diseases in both plants and animals. There are other phosphatic fertilisers available but their use is relatively rare; steam bone flour has the special property of drying a mixture, and is often incorporated to the extent of 1–2 cwt. per ton, to overcome stickiness in the other constituents.

The following are the general effects of phosphatic fertilisers:

(*a*) *They advance maturity*. This effect is a very definite one and is often seen in the case of barley. It is presumably because of this effect that phosphatic manures tend to raise the sugar content of sugar beet. With market-garden crops the effect may be very valuable, as has been shown experimentally to be the case with brussels sprouts and lettuces.

(*b*) *They stimulate root growth.* For this reason they are particularly important for such weak-rooted crops as barley, swedes, turnips and potatoes.

(*c*) *They are reputed to counteract the effect of excessive nitrogenous manuring.* This is very generally believed and sounds reasonable, because an unbalance of plant food may be expected to swing over the metabolism of the plant; thus to bring the other plant foods up to the same level should tend towards a normal plant. It must be confessed, however, that experiments have signally failed to establish the point; it has been conclusively shown that phosphatic manures do not help to check the lodging of cereals caused by excessive nitrogen, and there is little evidence that they counteract the diseases to which plants are prone when they receive much nitrogen.

(*d*) *They are especially beneficial to leguminous plants.* There is no doubt about the truth of this, and the explanation may be that these plants, being provided by bacteria with a sufficiency of nitrogen, are more liable than others to suffer from phosphatic deficiency as a limiting factor. It has been shown that phosphates tend to force the symbiotic bacteria into the motile phase of their life cycle, and this may explain, in part, the particular usefulness of phosphates to legumes, especially during the establishment of the crop.

(*e*) *They act as an insurance.* This is commonly quoted, and particularly applicable to root crops. In dry years phosphatic manuring is very beneficial to swedes, whereas in wet years it is relatively ineffective; this may be ascribed to its action in stimulating root growth.

Excess of phosphate is not washed out of the soil, and is not in general liable to lead to any harmful results. It was believed that excess of phosphate was never harmful, but work in Devon and Cornwall has shown that it may be so if it is associated with a deficiency of potash. Examination of the soils on which crop failures in those two counties occurred showed that they were characterised by a high ratio between the available phosphate and the exchangeable potash; where this ratio approached 10:1 danger signs began to appear.

Wheat and oats are frequently grown without any phosphatic manure, especially where the soil is light; if little of this manure has been applied to the field in the last year or two a dressing of 0·30 to 0·50 cwt. per acre of phosphate is advisable. Barley receives phosphates more often than the other cereals, because phosphate appears to make the barley ripen not only earlier but more evenly; a dressing of 0·50 cwt. of phosphate per acre is not uncommon, unless the barley follows roots which were generously dressed. In some cases the cereals are undersown with seeds and then the phosphatic dressing should be much heavier; 0·80 to 1·50 cwt. as superphosphate, or more often, basic slag, are then commonly given, but the manure is intended for the seeds rather than for the cereal.

Pulse crops should be generously dressed with phosphate, in accordance with the statement made above that they respond well. Beans should receive 0·80 to 1·20 cwt. as superphosphate or basic slag (usually they get rather less of the former, because of its greater availability and higher cost) and peas should receive a similar dressing, though in practice they usually get less.

Of the root crops, turnips and swedes are the ones for which phosphate is most important, because of their weak rooting habit. The dressing should not be less than 0·50, and may very well be 1·00 cwt. per acre, and if the soil is low in lime reserve the manure should be basic slag rather than superphosphate, because of the danger of finger-and-toe disease. Sugar beet normally receives a dressing of 0·30 to 0·60 cwt. as basic slag or superphosphate, though experiments have been by no means unanimous in supporting the practice; the slight tendency to raise sugar content may be held to justify a moderate dressing. With mangolds there is very little evidence that phosphatic manuring, on the majority of soils, increases yield; on the other hand, the mangold is a slow-maturing plant so that phosphate may be helpful in raising the dry-matter content, and a dressing of 0·30 to 0·60 cwt. per acre is usually advised. For both sugar beet and mangolds phosphate is unimportant relative to nitrogen and potash, except on the soils which analysis shows to be seriously deficient in phosphate; on such land phosphatic manuring is absolutely essential as without it there may be no crop worth harvesting. With green crops the belief was that it was only nitrogen that was required, but experiments have shown that a more balanced manuring produces a better yield, and 0·40 cwt. of phosphate per acre may be advised. Potatoes respond fairly well to phosphatic manures, and this is ascribed to the beneficial action on their roots; it has been argued that the rate of dressing should not be high, as otherwise the tendency to maturity leads to smaller tubers, but the argument is contrary to such experimental results as have been obtained. On most soils a dressing of 0·50 to 0·80 cwt. per acre would be considered ample, but on fens and silts 1·00 cwt. is deemed moderate.

Leys should receive a generous dressing of phosphate and, as mentioned above, the best practice is to apply up to 1·50 cwt. per acre to the corn crop in which they are undersown. For established grassland the usual recommendation is 1·50 cwt. per acre every four or five years, though after one such dressing many farmers only give half the amount on succeeding occasions. Basic slag is the form most favoured, but other forms may be used—superphosphate on soils with a high lime reserve and mineral phosphate on soils low in lime and situated in wet districts. Many poor grass fields on heavy land may be vastly improved by phosphatic dressing; the immediate effect is an astonishing spread of wild white clover, and this induces more luxuriant growth of the grasses, so that the feeding value of the pasture is raised to a much higher level.

It is impossible to give any idea of the sort of return to be expected from phosphatic fertilisers. It is certainly true that in many cases no return is obtained at all, but there are other cases where the success of the crop depends on phosphatic manuring. It is only experiments on the actual farm, or district, which can tell whether generosity or parsimony is advisable in a particular case, and usually such experience is not available; the reasonable line to take is to give moderate dressings, and those suggested above for the various crops may be regarded as moderate.

POTASSIC FERTILISERS

The main source of potash is the soil, and in the case of clay this usually provides an abundant supply, because the minerals from which clay is formed are potassic ones; thus their weathering naturally provides potash in amount which, on most clay soils and for most crops, are adequate. This, however, should not be assumed, because there are cases where heavy soils are deficient in potash and on them potassic manuring brings a rich return; during the War of 1939–45 the value of potash on a variety of soils, including clays, was repeatedly demonstrated and an occasional check by an advisory chemist is desirable on all farms. In general farming the amount of potash in the soil is maintained largely by the application of farmyard manure which contains much, whilst wood ashes are also rich in this plant food; salt has also been used as an indirect source of potash, which it renders available by displacement from the clay itself, but this is not to be highly recommended because sodium clay, which is very sticky, is formed. The two high-grade potassic fertilisers are sulphate and muriate of potash. The former, although containing slightly less potash, is rather higher in price than the latter, a relative position it is able to maintain by reason of its special value for potatoes, and for market-garden crops. At the other extreme is kainit, of which about 4 cwt. is needed to supply the amount of potash contained in 1 cwt. of muriate of potash; intermediary between the high- and the low-grade potash manures are potash salts, which are normally available in two grades, containing 20 and 30 per cent of potash. Apart from sulphate of potash these manures are sold at prices roughly proportional to their potash content; that being the case many farmers use only the high-grade ones, to avoid carting.

The general effects of potassic manures are very imperfectly understood, but the following are regarded as the main ones:

(*a*) *Potash is necessary for the formation of carbohydrate.* In this connection it is significant that the crop—potatoes—for which potassic manure is most necessary is the one which produces the greatest weight of starch per acre. Probably the common farm crop next most responsive to potash is sugar beet, which, again, produces a large weight of carbohydrate per acre.

(*b*) *Potash increases resistance to disease.* This effect appears to be fairly well established and to be manifested by a number of crops, both in regard to fungoid diseases and to insect pests. This effect is very variable. In the case of fungoid diseases potash does not stop infection of the plant but tends to check multiplication of the pathogen. An exception is club root of brassicae; susceptibility to this disease is increased by potassic manuring.

(*c*) *Potash is reputed to counteract the ill effects of excess of nitrogen.* The principle of this is the same as for phosphate, that is, that by supplying these other two plant foods the metabolism of the plant is prevented from being swung over by excessive nitrogenous supplies; but the evidence in support of the principle appears to be lacking. It has been claimed for potash that it is particularly useful in this respect with cereals, tending to make a strong standing straw; this is based on the fact that straw itself has a high potash content, but experiments have completely failed to demonstrate that potassic manuring has any effect in making straw stronger.

As far as is known there is no danger from excessive potassic manuring, and little is leached from the soil though potash may be immobilised in the soil. As stated above, no farmer should assume that his land is rich in potash, but many on heavy land, especially with regular application of farmyard manure, will find that potassic fertilisers need hardly ever be used, except for potatoes; it is the lighter soils of the country and particularly the chalks that respond best to potash.

Wheat and oats should never receive potash manures on heavy land and rarely get them on light land, though in the latter case compound fertilisers containing this plant food are often used. With barley the case for potassic manuring is stronger, partly because barley is grown more widely on light land, and partly because potash is believed to produce a brighter coloured sample; too much faith must not be placed on the latter point.

Beans are usually grown on heavy land and receive no potassic manures. Where farmyard manure is applied there is probably nothing to be gained by applying potash, but in the absence of farmyard manure beans respond generously to this plant food; 0·50 to 1·00 cwt. of potash should certainly be applied where no dung is given. Peas also respond to potash and should receive from 0·75 to 1·50 cwt. of potash per acre.

Of the root crops swedes, turnips, kohlrabi and cabbage frequently receive no potassic fertiliser if farmyard manure is applied; in the absence of farmyard manure, and on light land, dressings of about 0·50 cwt. per acre of potash are commonly given. Sugar beet and mangolds should receive rather more potash, and a common dressing is 1·50 cwt. of potash, in addition to farmyard manure. Kainit is frequently advised as the potash manure for these crops, the basis for the advice being that sugar beet and mangolds have been developed from maritime wild plants; from this it is inferred that they will utilise the variety of non-potassic salts included in kainit. If kainit is used for these crops it should be applied to the land some time before sowing the seed, as the large amount of salts contained in a dressing of 4 or 5 cwt. of kainit per acre may, in a dry time, make the soil solution too concentrated; the effect of this is that water will not soak through the fruit coat to the seed, and germination will be delayed. Generally an interval of six weeks or so between applying the kainit and sowing the seed is sufficient, but clearly the safety limit is determined by the rainfall before and immediately after sowing. Common salt may be used for sugar beet or mangolds in place of any potassic manure; the usual dressing is 5 cwt. per acre (applied, as with kainit, well ahead of drilling) and in experiments this has produced responses equal to that from 0·75 cwt. of potash. Common salt may also be used (at 2–3 cwt. per acre) in place of potash for barley and there are presumably other crops for which the substitution is permissible when potash manures are very expensive or in short supply; salt should not be used for potatoes, since it lowers their quality, nor is it desirable on clay soil which it tends to make stickier.

For potatoes potash is very necessary, and a reasonable dressing is 1·50 cwt. per acre. The chloride radicle is held to have a deleterious effect on the cooking quality of potatoes—to make them watery—and as potash salts and kainit and, of course, muriate of potash, contain a high proportion of this radicle it is unwise to use any of these manures for potatoes; thus it is that sulphate of potash has, as it were, a closed market for this crop. Experimental tests have shown that the harmful effect on quality does, in some cases, ensue when other potash manures are applied, but this is apparently only true on some soils; unfortunately, the soils on which it happens are not yet definable, so that it remains the only safe recommendation that potatoes should receive no other potassic manures but sulphate of potash. During the War of 1939–45 muriate of potash was perforce used for potatoes on all types of soil, because the small available supplies of sulphate of potash were reserved for tomatoes. The cooking quality of some potatoes probably suffered somewhat, but the country was in dire need of the crop, for which there was a guaranteed sale, and slight variations in quality passed unnoticed. Potato growers have not in general returned to the pre-war insistence on sulphate of potash as the only suitable potassic fertiliser for potatoes; indeed, a common practice is to use a proprietary 'potato' compound manure in which the radicle attached to the potash constituent is not declared.

Leys and grassland receive very little potassic manure. On light land, leys would often respond, but the general practice is to apply potash to the other crops in the rotation, particularly to the root crops. Permanent pasture undoubtedly responds well to potash on light land, and there has been a number of cases where great gain has followed its use; in those cases great success has followed the application of potassic and phosphatic fertilisers at the same time, despite the failure of each type when used separately. Grass occupies a small proportion of light-land farms, and on heavy land response is very rarely obtained from potassic manures. Lucerne leys are common on light land, where they respond well to potash; a dressing of 1·00 cwt. per acre of potash every other year is considered good practice for the crop.

COMPLETE MANURING

The best manuring for a particular crop depends on many things, and it is only with the greatest reserve that typical dressings are given. The factors that must be considered have

TABLE III. TYPICAL MANURIAL DRESSINGS

Crop	Preceding crop	Soil type	Dressings per acre				Notes
			tons F.Y.M.	cwt. nitrogen	cwt. phosphate	cwt. potash	
Wheat	Corn	Heavy	—	0·60	0·40	—	F.Y.M. if available. Most of nitrogen in spring
Wheat	Seeds	Loam	—	0·40	0·30	—	F.Y.M. commonly applied on lighter land All the nitrogen in spring
Wheat	Beans, potatoes, fallow, long ley	Loam	—	0·40	—	—	All the nitrogen in spring
Barley	Corn	Light	—	0·50	0·40	0·40	—
Barley	Folded roots	Light	—	—	—	—	0·40 nitrogen if winter is wet
Barley	Carted roots	Loam	—	0·40	0·30	0·30	—
Oats	Corn	Loam	—	0·40	0·40	0·25	F.Y.M. if available
Rye	Corn	Sand	—	0·50	0·40	0·50	F.Y.M. if available. As catch crop receives nothing
Beans	Corn	Heavy	12	—	1·00	—	Slag often used instead of superphosphate (0·50 cwt. potash if no F.Y.M.).
Beans	Long ley	Heavy	—	—	0·80	0·50	—
Peas	Corn	Light	—	—	0·60	0·50	F.Y.M. if available and land poor
Sugar beet	Corn	Loam	12	0·80	0·40	1·00	5 cwt. salt can displace half of potash
Mangolds	Corn	Loam	15	0·80	0·40	1·00	5 cwt. salt can displace half of potash
Swedes, turnips	Corn	Light	—	0·50	1·00	0·50	Or 12 tons F.Y.M. and 0·80 cwt. phosphate
Kale	Corn	Loam	12	1·20	0·40	0·50	F.Y.M. often omitted
Kohlrabi	Corn	Loam	12	0·80	0·40	0·50	—
Potatoes	Corn	Loam	15	0·80	0·80	1·50	—
Spring cabbages	Corn	Loam	15	1·00	0·40	0·50	Nitrogen applied between rows in spring
Rape	Corn	Light	—	0·60	0·40	0·50	Very often receives nothing
Mustard	Corn	Light	—	—	—	—	If for seed general dressing as shown for rape
Maize	Corn	Loam	12	0·70	0·40	0·50	—
Buckwheat	Corn	Fen	—	—	0·30	0·25	—
Linseed	Roots	Loam	—	0·40	0·40	0·75	—
Tares	Corn	Loam	—	—	0·80	0·50	—
Short and long leys	Corn	Loam	—	—	1·00	—	Often slag; applied to crop in which undersown

already been discussed, and Table III should be read with a clear realisation of all the various modifications that special circumstances may dictate. In order to avoid a very lengthy list of cases, dressings in most instances are only given for one soil type, which has, as far as possible, been chosen as the main type on which the crop is grown. Heavy land may require rather more generous manuring with phosphate, and light land with potash, than is shown in the table, which is, of course, very far from exhaustive, but which, it is hoped, covers sufficient cases to serve as a general guide.

The dressings given in the table may be regarded as moderate ones, but they are, nevertheless, higher than the average in general practice, because a few farmers still use no chemical fertiliser whilst a large number are very niggardly in their applications. In the year July 1955 to June 1956 the farmers of the United Kingdom spent on chemical fertilisers some £75 million, less nearly £18 million Exchequer subsidy; thus their net expenditure amounted to nearly £2 an acre of crops and grassland. Expenditure is, on average, still far below the economic optimum and this is particularly true in regard to grassland; farmers in arable districts, growing a high proportion of cash crops, usually spend upwards of £5 an acre of their farms on fertilisers. Surveys of fertiliser practice are carried out jointly by the National Agricultural Advisory Service and Rothamsted Experimental Station at intervals of about two years. Each Survey is restricted to certain selected counties or parts of counties, the latest published one being conducted in 1954 in twelve districts of England and Wales. In each district a number of farmers, selected at random, are visited and the amounts of fertiliser they have actually applied to their fields are ascertained. The collated data provide a picture of what is happening and, since some districts are included at successive Surveys, of the changes which are occurring. They show wide variation in practice between and within districts and also a general tendency for increasing fertiliser application; in nine districts surveyed in 1951 and 1954 there was in the three-year period an increase (on average of all crops and grassland) of 30 per cent in nitrogen usage, with corresponding percentages of 10 and 35 for phosphate and potash.

The relative popularity of the different fertilisers is shown by the following figures (thousands of tons) of deliveries as given in the Trade Returns for 1955–6:

Sulphate of ammonia	222
Other nitrogenous fertilisers	439
Superphosphate	135
Rock phosphate	37
Basic slag	712
Other phosphatic fertilisers	34
Potassic fertilisers	43
Total of straight fertilisers	1622
Compounds	2302
All chemical fertilisers	3924

These fertilisers contained the following amounts (thousands of tons) of plant nutrients:

Nitrogen	291
Phosphate	386
Potash	305

In weight the compounds amounted to 59 per cent of the total but in cost the percentage was 71. The rise in the importance of compound fertilisers has been continuing for a quarter of a century, having been stimulated by the introduction of ammonium phosphate, containing two of the plant foods in soluble form, and by the development of methods of granulation by which potash may also be included, to form the 'concentrated complete fertilisers', each granule of which contains all three of the plant foods. This has been claimed as a great advantage because it ensures that the nutrients are necessarily presented together to the plant roots. The claim has not been substantiated experimentally; the general conclusion to be drawn from trials is that these compounds are just about equal in effect to equivalent mixtures of straight manures. What cannot be disputed is that the concentrated nature of the compounds economises in transport and in labour of application, and the further fact that their condition is excellent.

The advantages of buying fertilisers mixed ready for application are undoubtedly considerable, and many farmers hold that they outweigh the disadvantages, of which the main one is lack of flexibility; if all the relevant conditions are to be taken into account there should be a separate prescription for each field, and this can only be followed by the farmer who mixes his own manures as required. A manure merchant is equipped with machines that will give a more thorough mixing than can be obtained on the farm; but if the sacks of manure are emptied in a heap, and the heap is turned two or three times with a shovel, the diffusion of the ingredients will probably be sufficient for all practical purposes. There are, of course, some fertilisers which should not be mixed, and a short list of the commoner cases may be given. The following should never be mixed, though application to the land with only a few days' interval between them is safe:

Sulphate of ammonia Nitro-chalk Superphosphate	with	Nitrate of lime Calcium cyanamide Basic slag

The following may be mixed, but application to the land should be within a day or two, because of changes which occur, and of the tendency to set:

Sulphate of ammonia Superphosphate	with	Mineral phosphate
Nitro-chalk	with	Superphosphate Kainit Potash salts
Nitrate of soda	with	Nitrate of lime Superphosphate
Nitrate of lime	with	Calcium cyanamide Mineral phosphate Basic slag All potash manures
Calcium cyanamide	with	All potash manures
Basic slag	with	Kainit Potash salts

It is regrettable, but true, that many farmers still have little knowledge of chemical fertilisers and of their possibilities and limitations, and for such farmers proprietary compounds are certainly advisable. The manure merchant is fully aware of all the pitfalls in mixing fertilisers, and furthermore, his product must be in good condition to sell well; he achieves condition by the admixture of 5–10 per cent of steamed bone flour, or by allowing the heap a long time to set and then regrinding. Condition in a fertiliser is always important and becomes supremely so when it is to be used in a combine drill; sticky, lumpy manure will not run well in the drill and causes great waste of time which may be very serious if work is being pressed on in a fleeting spell of favourable weather. Straight manures are normally in good condition and easy to mix when they arrive on the farm, but if they have been stored for some time, piled five or six sacks high and not absolutely dry, they may set in the bags and then mixing involves much work; beating with the back of a shovel, rolling with a garden-roller and other devices have to be used to break down the lumps and it may occupy two men fully to keep the drill supplied. Where a farmer has good, dry storage available he can avoid these troubles and can also take advantage of the rebates which are offered by manufacturers for early delivery. These are offered, for straight and for compound fertilisers, to encourage farmers to take delivery at seasons other than the spring when the great rush of orders occurs; rebates may be as much as 30*s.* a ton for July or August delivery, diminishing to 5*s.* or 10*s.* by January. An advantage of compounds is that large firms employ agents qualified to advise the farmer on manurial problems, and that they normally have a range of mixtures to offer which, supplemented perhaps with straight nitrogenous fertiliser, will meet most conditions satisfactorily. There is, however, for the uninformed farmer one serious danger involved in buying proprietary compounds—that, to speak colloquially, he may be 'done'. It is, of course, true that the analysis of a fertiliser must be declared at the time of sale, but too often farmers take little notice of this safeguard. Yet it is a very simple matter to check the price. First it is necessary to calculate the unit value of each of the plant foods; this is the value of 1 per cent of a ton of the actual plant food, and is obtained by dividing the price per ton of a straight fertiliser by its percentage content of nutrient. At the time of writing chemical fertilisers are subsidised to the extent of 30 per cent of the cost of nitrogen and 45 per cent of the cost of phosphate, with nothing for potash, all of which is imported. Quotations for fertilisers are for gross costs but they are always accompanied by statements of the amounts by which the actual costs to farmers are abated by subsidy. It will be best to work with net costs. Thus:

If sulphate of ammonia, containing 20·8 per cent of nitrogen costs £14. 1*s*. 0*d*. per ton, unit value of nitrogen is

$$\frac{\pounds 14.\ 1s.\ 0d.}{20{\cdot}8} = 13s.\ 6d.$$

If superphosphate, containing 18 per cent of phosphate costs £8. 1*s*. per ton, unit value of phosphate is

$$\frac{\pounds 8.\ 1s.\ 0d.}{18} = 8s.\ 11d.$$

If muriate of potash, containing 50 per cent of potash, costs £17. 2*s*. 6*d*. per ton, unit value of potash is

$$\frac{\pounds 17.\ 2s.\ 6d.}{50} = 6s.\ 10d.$$

The value of a compound fertiliser may then be calculated as follows:

Analysis of compound	Unit value		£	*s*.	*d*.
12 per cent nitrogen	13*s*. 6*d*.	12 × 13*s*. 6*d* =	8	2	0
15 per cent phosphate	8*s*. 11*d*.	15 × 8*s*. 11*d*. =	6	13	9
7 per cent potash	6*s*. 10*d*.	7 × 6*s*. 10*d*. =	2	7	10
		Value of 1 ton of compound =	£17	3	7

The manure merchant is entitled to add something (say £1 to £2 per ton) to the price for the services he renders, such as giving advice, thorough mixing, and producing a compound which will be convenient to handle; it is unfortunately true that many cases have come to notice where a compound manure has been sold at a greatly excessive price. All farmers should apply the above check to every quotation, substituting, of course, in the calculation the current prices for the straight manures.

It is convenient to work to a general plan in manuring, and one such may be given as an example. The following common heavy-land rotation might be accompanied by the manures shown:

Beans	12 tons farmyard manure, 1·00 cwt. phosphate
Wheat	0·40 cwt. nitrogen as nitro-chalk in late April
Barley	0·50 cwt. nitrogen, 1·00 cwt. phosphate as basic slag
Seeds	No manure
Wheat	12 tons farmyard manure, 0·40 cwt. nitrogen in late April
Oats	0·40 cwt. nitrogen, 0·45 cwt. phosphate

It will be realised that the generous dressing of basic slag to the barley crop is designed primarily for the benefit of the seeds of which this is the nurse crop. The farmyard manure produced on the farm would be more than sufficient to provide a dressing for one-sixth of the arable area, but hardly enough for two-sixths; accordingly some of the wheat following seeds would receive no farmyard manure, the fields in better heart than the average being the ones to suffer. Where wheat after seeds received no farmyard manure it should be combine drilled (see p. 92) with 0·20 cwt. of nitrogen and 0·30 cwt. of phosphate; it should still receive 0·40 cwt. of nitrogen between mid-April and mid-May. It will be noticed that no application of potassic manure is included, and this is in conformity with the fact that many heavy-land farmers use none of these manures. Soil samples should, however, be taken and analysed from time to time to check that the potash status of the land is being satisfactorily maintained. In any case the manurial plan must not be adhered to rigidly, and the farmer must ever be ready to modify it in accordance with the condition of particular fields (as evidenced by the crops they carry), or in the light of his own and his neighbours' experience.

It is not possible to draw up a neat balance-sheet, to show on the one side the plant foods added to the soil and on the other those removed by the crop, and to deduce from this the net gain or loss made by the soil in a season. Results would be vitiated by losses from leaching and by the locking up in the soil of plant nutrients whilst weathering effects may be acting in the reverse way and rendering unavailable plant foods available. Many chemical determinations have been carried out to calculate the balance for a crop and these show that the crop takes up but a small proportion of the foods supplied in manures. There would appear, therefore, to be considerable residuals from fertiliser applications which should benefit succeeding crops.

These residuals become important when a farm changes hands, since an outgoing tenant is entitled by law to compensation for the unexhausted value of improvements he has made and hence to residual fertility in the land from manures he has applied. As long ago as 1875 Lawes produced a table of residual values to guide valuers in their difficult task of assessing how much an ingoer should pay his predecessor under this head. The table has been revised from time to time, the latest revision being made by a Conference convened by the Ministry of Agriculture in 1946. For the full results of their labours readers must be referred to the *Report* of the Conference; only the briefest summary will be given here. In regard to fertilisers the *Report* recommends valuers to carry forward the following fractions of the cost:

Fertiliser	After 1 crop	After 2 crops	After 3 crops
Nitrogenous:			
Inorganic or dried blood	—	—	—
Organic (except dried blood)	$\frac{1}{2}$	$\frac{1}{4}$	—
Phosphate:			
Soluble phosphate	$\frac{2}{3}$	$\frac{1}{3}$	$\frac{1}{6}$
Insoluble phosphate	$\frac{1}{3}$	$\frac{1}{6}$	$\frac{1}{12}$
Total in bone products	$\frac{1}{2}$	$\frac{1}{4}$	$\frac{1}{8}$
Potassic	$\frac{1}{2}$	$\frac{1}{4}$	—

In regard to lime it is recommended that, provided the dressing was a proper one, seven-eighths of the cost should be charged after 1 year, six-eighths after 2 years and so on, the whole being wiped out in 8 years. Figures are also given for the residual values of feeding-stuffs consumed on the farm, with adjustments which should be made according to the type of animal which consumed them and to the way in which farmyard manure has been handled. When all the arithmetic has been done there remains much for the judgement of the valuer and this is well, because experiments designed to measure residual values of manures often fail even to detect them. The whole matter is obscure because despite these experiments practical opinion is unanimous that an outgoing tenant who has done his land well receives inadequate compensation, whereas one who has farmed his land out is not mulcted heavily enough.

TIME AND METHOD OF APPLICATION OF CHEMICAL MANURES

Earlier in this chapter the times when fertilisers may best be applied to the land have been mentioned. Nitrogenous manures may be applied as top dressings at various stages of growth, but the other fertilisers and, in fact, much of the nitrogen as well, are usually applied to the seedbed. Potassic manures may prove very effective where deficiency symptoms are observed quite late in growth as, for instance, in April on autumn-sown wheat, but they are best applied at drilling time; the latter is certainly the correct time to apply phosphatic manures which help greatly in plant establishment. Seedbed application needs a word of explanation. Just before the seed is drilled—whether it be corn or roots—some final cultivations are necessary to prepare the final tilth, and the expression 'applied to the seedbed' should be taken to denote application just before the final cultivations, which will serve to work the manure into the ground. This is the usual time to apply all the fertiliser, except for such nitrogenous manure as it is decided to reserve for use as a top dressing. In the case of root crops it is good practice to work the land some weeks before drilling, and there is something to be said for applying the manure to the winter furrow before this initial working; it will then be worked into the land more deeply, but the nitrogenous part of it may suffer leaching if very wet weather follows. The reader is reminded that when kainit or salt is being used, for sugar beet or mangolds, it is wise to apply that manure some six weeks ahead of drilling, because of the danger of reducing plant establishment. With potatoes a common practice is to apply the fertiliser over the field when it is first ridged up, immediately before the tubers are planted; thus, when the ridges are split to cover the tubers, the manure is covered in, in the new ridges.

It is commonly asserted that even spreading of chemical fertilisers is essential for the best results, and that this is of the greatest importance where concentrated fertilisers are being used. Numerous tests of manure drills for evenness of sowing have been conducted, but the testers have concentrated on the distribution of the fertiliser over the surface of the ground, rather than on crop yields; it is to be presumed that, up to a point, good distribution is desirable, but it is only presumption, and, in any case, the required point is as yet undefined. The whole question is of some importance because engineers have found it difficult to produce the ideal manure distributor. Such a machine would sow anything from 1 cwt. to 1 ton per acre of all sorts of fertilisers, some highly corrosive, varying widely in condition, size of particle and weight per unit volume; these are formidable requirements, and perfection in distribution, towards which most efforts are directed, may not be so important as to justify sacrifice on the more practical points regarding general utility. There is no doubt, of course, that the manures should be spread reasonably evenly over the width of working, but most machines will do this satisfactorily, because it is rare to see uneven growth in the crop occasioned by more distribution from one part of the machine than from another.

It is on this last point that broadcasting manure by hand usually fails. The tendency is for the man to throw more direct to his front than to either side, and it is not at all infrequent to see, by the more luxuriant growth subsequently, each track across the field of the broadcaster. In favour of broadcasting by hand is the fact that it is cheap and that no outlay for a distributor is occasioned; furthermore it can be done without damaging the crops when they are up, and is therefore the best method of applying nitrogenous manure to wheat in late spring. But with some fertilisers—calcium cyanamide, basic slag and, to a less extent, superphosphate—the work is very uncomfortable, particularly if the day is at all windy. The method is still used but good work requires much practice, and the number of men with the necessary experience is very limited.

There are many different types of manure distributor, and it is not proposed to enter here into details of their construction. An early type consisted of a circular hopper from which the manure fell on to revolving disks which threw it out; this was like hand broadcasting in giving a very uneven distribution, particularly on a windy day, and the type is now very rarely seen. The common form of distributor consists of a box, containing the manure, extending the full width of the machine, which is usually 8–10 ft.; distribution may be from the bottom or from the top of the box. Bottom delivery has the advantage that the manure has less distance to fall to the ground and is therefore not so liable to be blown about; various devices, such as an endless chain, revolving cylinder or reciprocating bars, are used for the actual delivery. The drawback to bottom delivery is that the manure in the box may 'bridge' and not fall down to the delivery mechanism; this was much more likely to happen before the introduction of the dry type of superphosphate, and should cause no trouble with well-conditioned, compound fertilisers. The top delivery type overcomes any difficulty from 'bridging', the principle being that the bottom of the manure box is gradually wound up as the distributor proceeds across the field; thus the manure is constantly pushed above the top of the box, where a revolving spiked bar scrapes it over the back of the box. In general one horse is all that is required to pull a manure distributor, though if the ground is very rough, or soft, two may be necessary; with a tractor a battery of three distributors —the centre one being advanced so that the wheels may clear— is not infrequently used.

There is another method of applying chemical fertilisers. This is to drill the manure with the seed, using what is generally called a combine drill. The practice is by no means new, particularly in some parts of the country, but the modern method differs very materially from the old one. The practice was very general in the fens, where the manure commonly applied was superphosphate and that manure used to be very sticky; to make it run better it was mixed with ashes, and so the drill became known as a 'compost' or 'compass' drill. The drill carried two separate boxes, a front one for the manure and a rear one for the seed, and two separate sets of coulters, the rear ones running in the tracks of the front ones; thus the front coulters sowed manure, and some powdering of earth fell over this before the rear ones deposited the seed. Combine drilling in this way was undoubtedly quite safe, but in most modern machines the manure and seed travel together down the same coulters, and are in some cases actually mixed together in the same box. Furthermore, instead of 'diluted' superphosphate, concentrated fertilisers containing nitrogen and potash are used. In these circumstances there is some danger that plant establishment may be seriously reduced. The danger is that when the seed lies surrounded by concentrated fertilisers the soil moisture, in its immediate vicinity, may become so concentrated in salt content that water will not pass into the seed by osmosis. Clearly the danger only arises in a dry period, but cases have been seen where germination has been delayed by lack of rain for a month or more. This is particularly serious in a crop like sugar beet, as the few plants which appear may cause the farmer to think that he has a failure and to redrill, or alternatively he may be satisfied and single the crop, only to find that this laborious work must be done again when rain eventually comes. Experiments have shown that it is the nitrogenous and potassic, rather than the phosphatic, fertilisers which are responsible for the effect, and consequently the safety of the old practice, proved by years of experience, cannot be argued in favour of the new methods, If a sufficiency of moisture is present in the soil at the time of drilling, or if adequate rain falls soon afterwards, no reduction in germination occurs, and there is ample evidence that manure drilled with the seed then produces a greater increase in yield than the same amount of manure sown on the surface of the land. In many cases the added efficiency, as compared to broadcasting the fertiliser, is very great and can easily be seen throughout the growth period of the crop. The method is particularly useful where fertiliser is in short supply, as a small dressing 'down the spouts' produces nearly as great a response as double the amount broadcast. During the War of 1939–45 many combine drills were imported and this method of applying fertilisers gained greatly in favour; at the present time a much wider spread of combine drilling is only prevented by the shortage of that type of drill. In Great Britain there were, in 1954, some 37,000 combine drills as compared to 97,000 ordinary drills. One interesting case has been reported, in which manure drilled with barley produced a much higher yield than manure broadcast; this crop was severely attacked by wireworm, which were apparently greatly inconvenienced in the rows where manure was drilled, but the combine drill cannot be described as a reliable way of controlling wireworm. Combine drilling is well established and may be regarded as profitable and safe with cereals but the risk of impairing germination is too great with pulse or root crops or with the small seeds of grasses and clovers.

For these, 'band' sowing would seem to be the method of the future. In this method the drill has separate boxes for the seed and the fertiliser, each seed coulter having a fertiliser coulter running about 2 in. to one side of it. Thus the fertiliser is sown in a band within easy reach of the first rootlets and yet far enough away to avoid any harm to germination. Many experiments have shown that the method gives practically the same benefits as combine drilling without any of its risks. Unfortunately it doubles the number of coulters in the drill and presents mechanical difficulties because if coulters are too close together blockages may occur either with clods or with rubbish. The coulters of an ordinary drill are staggered to give the machine better chance to clear itself but with narrow rows and double the number of coulters clearances must be less and it is difficult to make a machine work satisfactorily on any but the cleanest and finest of seedbeds.

A similar procedure is indicated for planting potatoes. It has been shown that 7 cwt. of fertiliser applied over the open ridges before planting gives the same yield increment as 10 cwt. broadcast over the field before the ridges are opened. It is now quite common for potatoes to be planted mechanically, the machines for this opening the ridges, dropping the potatoes and closing the ridges, all at one passage. Some machines carry a box for fertiliser which is delivered down the coulters through which the seed tubers drop, the latter receiving a dusting of fertiliser as they fall. In the great majority of cases this works admirably, even if as much as 12 cwt. of fertiliser is sown, but in some cases the young potato sprouts are affected, with consequent delay in growth; as with combine drilling, the risk is greatest when planting conditions are dry. Engineers will probably devise a means of band sowing of fertiliser for potato planters.

Chemical fertiliser that is spread on the surface is not worked far into the land by harrow tines, and though nitrogenous manure may be washed down into the soil rapidly in solution, phosphatic and potassic manures are slow in this descent. It has been pointed out that these last remain largely in the surface inch or two for some weeks or months, and that plant roots obtain their supplies of nutrients more from the deeper than from the surface layers; this is particularly true of sugar beet, most of whose 'feeding' roots are between the depths of 6 and 18 in. The suggestion has been made, therefore, that since they are not likely to be washed out of the soil, the phosphate and potash for sugar beet should be applied in the previous autumn and ploughed in with the farmyard manure. As yet few experiments on the subject have been completed, but the present indications are moderately favourable, and it may be that further enquiry will show that at least a proportion of these two types of fertilisers should be applied in autumn on certain soils; all the nitrogenous fertiliser should, of course, be reserved for the spring, owing to the danger of loss by leaching during a wet winter.

GREEN MANURING

Green manuring is the practice of growing a crop for the purpose of ploughing it into the ground, to increase the amount of organic matter in the soil. Green manuring should not be confused with green soiling, a misnomer applied to the case where a green crop is cut as required and carted to livestock in yards or at grass. The value of organic matter in the soil, both on heavy and on light land, is fully recognised, but a green crop appears to be a very poor way of raising organic matter content; the straw contained in farmyard manure is a much more lasting source. There are farms, however, which are very short of farmyard manure, which is a commodity becoming ever more difficult to purchase, and then green manuring provides a rather unsatisfactory substitute. This method of maintaining the humus content of the soil may also be adopted for outlying fields, to which the carting of farmyard manure would be expensive. Experiments at Woburn and elsewhere have not shown green manuring in a very favourable light, and it certainly seems that the benefits accruing from it are insufficient to justify the sacrifice of a crop. This view conforms with general practice, in which crops grown for green manuring are usually in the nature of catch crops.

Mustard is frequently grown for green manuring, particularly on light soils. Its quick growth, enabling it to be grown after a fallow, and the fact that it produces a large amount of material for ploughing in, both make it a suitable crop for the purpose.

The difficulty is to get it all covered up with the plough, but a good man usually leaves little green showing, particularly if the crop has been rolled first; the rolling must be done in conformity with the plough ridges, so that the mustard is always leaning in the direction in which the plough travels, and the stems must not be broken off or they will be pushed along in front of the plough. It might be thought more reasonable to grow a leguminous crop for green manuring, so that the soil may be especially enriched with nitrogen, and tares suggests itself as a bulky crop. A long series of experiments at Woburn, however, showed that tares were less successful than mustard (which was, itself, rather ineffective); the explanation suggested was that the tares decomposed too quickly after ploughing in, and that the nitrogen was leached from the soil before the succeeding crop could utilise it. On heavier land the leaching would not be so serious and then tares might compare more favourably, but green manuring is not common on heavy land. On very light land inclined to be acid lupins have been used with some success for green manuring. Red clover is a crop very frequently used for green manuring; where it is intended to break a seeds ley after it has been cut for hay in June, sufficient time for a second growth is usually allowed before the field is broken up. In some districts red clover is grown solely for green manuring, the crop being undersown in a corn crop and allowed about two months' growth after the latter is harvested; the ploughing in of a young seeds ley in autumn is considered to be a fine preparation for potatoes in the following year. Trefoil is probably preferable to red clover for this purpose as the seed is usually cheap and trefoil is not subject to stem rot, which might be encouraged by the frequent growing of red clover. Some farmers who keep few livestock are attempting to maintain the fertility of their land by undersowing all their corn crops with trefoil for ploughing down in the following autumn. Various views are expressed as to the efficacy of the practice but as yet no satisfactory experimental evidence on the point is forthcoming. There are a few other cases where green crops are ploughed into the ground, but green manuring has never attained widespread popularity in this country; although there are to be found some farmers who are enthusiastic over the benefits obtained from green manuring the general practical view is that its utility is not great, and this is supported by the available experimental evidence.

The ploughing in of a turf, whether it be of a ley or of permanent pasture, is, of course, very different from green manuring. With the latter what is ploughed in is a young luscious crop which has had little time to grow much root, whilst with the former there is an accumulation of root growth together with withered remains of aerial growth which have been trodden into the surface. The grass turf contains long-lasting organic matter which with the droppings of the animals that have grazed it will benefit the field for several years. A ploughed-in turf may do the land more good than a full dressing of farmyard manure but no comparable benefit can be claimed for green manuring.

The subject of manuring is a very vast one in which much remains to be discovered. The decision as to what dressing should be applied to a particular field is still, and probably will be for some time, largely a matter of guesswork, and the lines along which a successful guess may be sought have been indicated. It is best to work to a general plan, carefully considered in relation to the rotation, but to be ever ready to modify the plan in the light of experience, as new knowledge is gained, or according to the special cirumstances of the fields. It can be said that experience has shown that it is better to spread the manuring, though not evenly, of course, over the rotation, rather than to apply all the fertilisers to one or two crops; it is probably true to say that in most other countries practice leans more towards applying all the plant nutrients evenly to every crop of the rotation than it does in this country. This leads to the final thought as to what is the object—is it to manure the crop or the field? The line taken in this chapter—the manuring of the crop rather than the field—would appear to be the right one to follow. In this view combine drilling is desirable since it evokes the same response in the crop drilled as would a larger dressing applied broadcast; the extra manure required by the latter method could only be justified by postulating valuable residuals. A farmer is wise who looks for the full return for his manures from the crop that gets them, but if he is generous over a series of years, and especially if a full head of stock gives him adequate supplies of farmyard manure, he will steadily raise the condition of his farm.

CHAPTER III

CLEANING

There are several possible definitions of a good farmer. He might be defined as a man who produces the maximum amount of food from his land, or as one who achieves financial success, but to an earlier generation a good farmer was primarily one who kept his land relatively free from weeds. And there is much to be said for this older definition, because it is generally true that a clean farmer produces heavy crops and is financially successful; it is certainly true that foul land betokens poor crops, and a general low level of farming. Farming involves unremitting war with weeds; there are general strategical principles and minor tactical moves in the struggle, and together these form a very important part of the art of husbandry.

A weed is frequently defined as a plant out of place, and it is essential that the reader should be quite clear from the outset that weeds are harmful. The various ways in which harm may be wrought by weeds may be classified as follows:

(1) They compete with the crop for space, light, plant food and moisture.

(2) They may smother the crop; that is, the useful plants may be completely shaded and dwarfed when the effect given above is accentuated. It is worth while noting that this effect may be reversed, and that a heavy crop will dominate, and tend to kill out, weeds.

(3) Some weeds wind themselves around the crop plants and, almost literally, strangle them; crop plants are frequently pulled over and deformed, bindweed being the commonest culprit.

(4) A few weeds are definite parasites, living on, and at the expense of, the crop plants. The common examples of this are dodder and broom rape on clover, and yellow rattle on grasses.

(5) The seeds of some weeds become mixed with corn and lower its value. The outstanding case in this connection is wild onion which imparts a very distinctive smell to corn.

(6) Some weeds are poisonous to stock, whilst others produce

a taint in the milk of cows that consume them. Many weeds are poisonous at some or all stages of growth, but fortunately livestock usually avoid them; that some unexplained deaths should be attributed to the consumption of poisonous weeds cannot be doubted, though, as a cause of mortality, this has probably been exaggerated.

(7) Some weeds provide alternative hosts for fungi and pests, and so perpetuate harmful species. From many possible examples might be selected charlock, as an alternative host for turnip-fly, and couch, as an alternative host for frit-fly.

Any one of these effects may be of the utmost gravity in a particular case, but in general it is the first on the above list which is of the greatest importance. The struggle for existence has continued for centuries, and in this struggle useful plants have been pampered relative to the weeds, whose extermination has always been the aim in agriculture. It is, therefore, scarcely a matter for wonder that the weed species which survive are very vigorous, as can be illustrated by the vast number of seeds produced by the dock, or the ability of a small piece of the rhizome of couch to establish a new plant. Thus it is that in the fight with weeds the crop, in the absence of outside assistance, loses. Examples of this are sadly frequent and total loss of crop sometimes occurs, whilst in a large proportion of crops reduction in yield, more or less serious, is occasioned; were it possible to calculate the actual loss in yield caused by weeds in this country the results would certainly be staggering.

GENERAL CLEANING METHODS

There is urgent need for more knowledge of the common weed species, of the phases of their growth when they are most vulnerable, and the operations to which they will most readily succumb. A field usually carries a large variety of different weeds, but in general there are one or two species which dominate, and cleaning is directed primarily at the prominent ones; if, therefore, operations could be rightly selected and accurately timed great efficiency would result. As it is, however, the necessary knowledge is lacking, and it is not possible to give precise instructions, complete with the experimental proof that they will be effective. Weed control is not, in fact, a science but rather a miscellany of possible steps, some of which should be taken regularly as a normal part of good practice, whilst some should be held in reserve until called for by the severity of infestation of special weed species.

The general methods of keeping land clean may be classified under six headings:

(*a*) *Cultivations.* Moving the soil is a very powerful method of weed control, though its efficiency varies much with the different species. When labour was cheaper the land was cultivated much more than it is at the present day, and the constant movement was the chief means of killing the weeds. The modern farmer must substitute intelligence for prodigality in cultivation, directing his operations so that they may take the greatest possible toll of weeds; this requires a detailed knowledge of the growth habits of individual weeds and the careful timing of operations. The tractor with its power of covering the ground quickly has greatly facilitated cleaning by cultivation but the efficiency of the work remains highly dependent on the weather; in wet conditions moving the land may do little good because most of the weeds, even annuals, then quickly re-establish themselves.

Of all weapons the plough is the most deadly in the fight against weeds. It not only undercuts them, but inverts and buries them, and growing annual weeds are completely exterminated; it should be noted, however, that good ploughing is necessary for this, as weeds which are not completely covered up may succeed in re-establishing themselves. Perennial weeds are capable of surviving the plough, but are weakened and will eventually be killed if the plough is used often enough; even couch may be eradicated on free-working soils by ploughing once every month from February to July. A single deep ploughing (to a depth of at least 14 inches) may be very effective in disposing of perennial weeds, especially couch. This requires a track-laying tractor and a deep-digger plough to turn all the surface in and leave the furrows flat. Farmers on good deep soil who are suitably equipped think little of an infestation of couch; they just plough it in deeply and forget about it. The harrow and cultivator are also useful weapons in the fight. The deeper the working of the land the more will the weeds be disturbed and the greater the setback they will suffer; young weed seedlings may sometimes be killed in prodigious numbers by a light seeds harrow, but older plants with fairly extensive root systems require a heavy harrow, and even then they do not succumb unless the operation be repeated one or more times. The actual seasons at which these cultivations are introduced will be dealt with later; here it is only desired to insist on the point that if they are intelligently directed, and if conditions are favourable, they will be very efficient in preventing the land from becoming foul.

(*b*) *Hoeing*. Hoeing is, of course, a form of cultivation, but since its main purpose is weed destruction it merits a separate heading. With perennial plants hoeing rarely has a lethal effect, but it may be useful in weakening the plant or in preventing seeding; creeping thistles can be greatly weakened, or even killed, by repeated hoeing of a root crop, whilst the hoeing of docks in spring, though not actually killing the plants out, will save much future trouble. Annual plants are usually killed by hoeing, as they are very vulnerable to cutting through the hypocotyl, or upper part of the root; they sometimes survive, however, in a wet time, if they are pulled up, rather than cut, by the hoe, and if they are immediately trodden into the ground again. It is the root crops, sown in rows wide apart, that give the greatest chance for hoeing, and they are therefore commonly referred to as cleaning crops. It must be pointed out that they are not necessarily cleaning crops, because the wide rows give great opportunity to weeds, and in early growth root crops do not compete successfully with weeds; but if full advantage is taken of the chance presented for hoeing, vast numbers of weeds can be killed before the rows over-top, subsequent to which a good crop will completely overshadow the ground and prevent weed growth. It will be appreciated that the date of sowing root crops contributes to their cleaning effect; unless they are sown early in spring there is sufficient time for many weeds to grow before drilling the crop, these being killed in the final preparation of the seedbed. Beans, also, are sown in rows wide apart and so offer good opportunities for the hoe; but they over-top much earlier in the year, and since they are normally grown on heavy land, there may be little chance for hoeing in a wet year. In most years there is at least one spell of dry weather in March or April, and then valuable cleaning can be done with the hoe in a bean crop. Peas also allow of some cleaning by the hoe, but the crop is not so helpful in cleaning the land as beans, because peas have very weak stems and lodge early in the season; though they prevent further work by sprawling all over the ground, the cover they give is not sufficient to smother weeds, which grow through them. In the past cereals were normally hoed by hand during the spring, but this practice is now practically extinct. When men are seen nowadays with hand hoes in corn they are usually 'chop hoeing', that is, they are walking through the crop chopping special weeds, usually thistles; the effect of this is valuable, but the cleaning is far less than that of a thorough hoeing, in which all the surface is moved, and all weeds are undercut. Thistles are also cut below the ground on grassland, but an ordinary hand hoe is rarely used because of the hardness of the ground; a spud or very light pick is the normal tool, and if the operation be done regularly once or twice a summer creeping thistles are kept well under control.

(*c*) *Smother crops*. If a crop does not fully cover the ground weeds will be only too ready to occupy any spare space. Where the crop has been a thick one, a corn stubble is usually free from weeds, but after a thin crop it is usually quite green; it may be a counsel of perfection, but it is very true that the best way to keep weeds down is to grow good crops. The farmer who is master of his job, well up with his work and generous in his manuring, reaps his reward not only in heavy yield but also in the ease with which he keeps his land clean; where heavy crops are grown the wastage involved in fallowing need never be incurred.

Crops vary much in their smothering properties. A silage mixture (beans, oats and tares), kale, rape, sugar beet, mangolds, and, to a lesser extent, other root crops and oats are all good smothering crops if they are thick, and get a clean start; thin corn crops, undersized roots, and, worst of all, linseed, are at the other extreme, and leave the land very foul.

(*d*) *Rotations*. Only a brief reference to this point will be made because it has been dealt with in Chapter I, but its great importance makes some slight repetition imperative. A succession of different crops ensures cultivation at different times in successive years, and this has a valuable cleaning effect. In addition a rotation brings cleaning and smother crops to all fields in due course.

(*e*) *Change of conditions.* By this is meant radical change so that the weed species which have established themselves in one environment shall find their circumstances completely altered. The change from arable to grass, or vice versa, is a case in point. Only few weed species thrive under both sets of conditions, and it is a telling argument in favour of alternate husbandry that it helps to keep the land clean. When a pasture becomes infested with watergrass, the quickest and best method of eliminating the weed is to break the land up and put it through an arable rotation. The annual weeds of arable land rarely grow in a grass field and a four or five years' ley gives sufficient time for most of their seeds to die. Even couch can be killed on many soils by putting the field to grass, because it will not long survive the firm conditions produced by heavy grazing on clay land.

Some weeds, e.g. spurrey and sorrel, are only found on acid land, of which they are valuable indicators. The only satisfactory method of eliminating them is to apply lime; when this is done they do not disappear forthwith but subsequently normal cultivations will steadily reduce them and there will be no fear of their gaining an ascendancy. Similarly species such as sedges, which indicate wet conditions, are eliminated by drainage.

(*f*) *Prevention of seed dispersal.* This is a many-sided principle requiring constant application. Clean seed is obviously desirable, but in many cases sufficient care to obtain it is not shown. If a million seeds are sown per acre (a modest rate for most crops) and the sample contains 1 per cent of weed seeds, the average number of the latter sown will be approximately 2 per sq. yd.; this is amply sufficient to provide a severe infestation of the crop. Weed seeds may be applied to the land in large quantities in farmyard manure, particularly where sweepings from hay lofts, etc., are thrown on to the dung heap; this was one of the reasons advanced in Chapter II in favour of carting farmyard manure to a large field heap, so that heat might develop in it. This gain, however, will be turned to serious loss if weeds are allowed to grow and to ripen their seeds on the field heap; ridiculous though it sounds, this happens quite frequently, the main weed concerned being fathen. Weed seeds are of course dispersed from field to field, or, worse still, from farm to farm, on implements, on the feet of men and stock and by birds, but in most of these cases there is little that can be done about it. Hedgerows are a fertile source of weed seeds, and it is well worth while to send men along them cutting down the weeds, in order to avoid future trouble; this is usually done about July, though some species have ripened their seed before then. A farmer has legal remedies against his neighbours if their land is dirty with some weeds, and can obtain an order requiring them to keep their weeds in check; in most cases farmers are loath to use these powers, however, as it does not conduce to a happy life to be continually taking one's neighbours to court.

SPECIAL CLEANING METHODS

Despite reasonably close attention to general cleaning methods the farmer often finds himself forced to adopt special measures against weeds; indeed this is so commonly the case that it is difficult to decide which methods should be termed general and which special. The first method described below is not directed against any one weed species, and as it occasions no dislocation of practice it might be held to qualify for inclusion as a general method; but it is not adopted as widely in practice as it might well be, and consequently it may be designated special.

STUBBLE CLEANING

The principle of stubble cleaning is to break the surface of the field immediately after harvest, so that weed seeds lying in the top inch or two of the soil may germinate; later in the autumn the young plants produced are killed by the normal ploughing which the field receives in preparation for the next crop. The essential points are that the breaking of the surface of the field should be shallow, so that weed seeds may not be buried too deeply for germination, and that a sufficient interval should be available for germination to occur. Therefore the operation should be carried out as early as possible after harvest, and it is generally restricted to fields that are to carry spring-sown crops, and are therefore not ploughed until well on into the autumn.

The special implement for the operation is called a broadshare, and it is a heavily built, rigid-tined cultivator carrying V-shaped shares, each about 18 in. wide. A width of working of 4–5 ft. is common, and for such an implement four horses would be necessary, and the area covered would be approximately 5 acres per day; but there is much to do on a farm directly after harvest, and horses could ill be spared for this slow and laborious work. Therefore it was unreasonable to urge farmers to use a broadshare before tractors became common, and the operation was rarely carried out. But the tractor is now almost universal, and is capable of working long hours when required. With the reduced amount of cultivation and less hoeing characteristic of the present time, there is every reason why the operation should now be general; nevertheless, it is far from general, few farmers being equipped with the necessary implement. Various other forms of implement are in use in different parts of the country. In Kent the old-fashioned broadshare consisted of a wooden gallows plough carrying one share; this required two horses and covered a width of about 18 in. Its work was excellent as it left the surface loose and in little ridges, giving conditions ideal for the encouragement of germination; but its work was slow and it is doubtful if it is ever used at the present time. On loose soils and in wet districts the form of broadshare favoured consists of a long bar which is dragged through the soil at a depth of about 2 in. This implement does good work in those districts, but will not face hard ground. On some soils the land is very hard after a dry summer, and even the heavy cultivator form of broadshare will not penetrate, though a disk harrow may do so. In such circumstances the surface loosening may be obtained with a shallow plough, followed by drag harrowing. This can be done quickly with a tractor, but involves more work than broadsharing and, since the furrow cannot be less than about 3 in. deep, the seeds are buried rather more deeply than is desirable. A method known as 'back striking' achieves good results on some soils. The field is ploughed with a double-furrow plough from which one mould board has been removed, and is then harrowed, across the furrow; this brings all the rubbish to the top where, given dry weather, it dies. The method may, therefore, be very efficient against perennial weeds, and after the operations have been carried out, the seeds of annual weeds will germinate in the loose soil produced.

The efficiency of stubble cleaning must depend to a large extent on the weather at the time. In a dry season the undercutting (that is separating them from their water supply) of the established weeds, which are weak and 'leggy' just after the corn crop has been removed, will lead to their rapid death; this effect may be valuable but should not be regarded as the chief purpose of stubble cleaning. Usually, even in a dry season, the weather breaks in early September, and then stubble cleaning may be very efficient in its main purpose of encouraging annual weeds to germinate. The operation should be performed as soon as possible after harvest—may even be done between rows of corn stooks on a wet day during harvest—and further germination is encouraged if the loosened soil is harrowed about a month later. Studies have shown that the normal winter ploughing—carried out in November or December—will achieve 100 per cent kill of the annual weed seedlings which are produced; incidentally nearly all the corn shed from the previous cereal crop will grow, and will be killed by the ploughing.

It must not be thought, however, that all annual weed seeds will be induced to grow, because the seeds of many species need a dormancy period before germination can take place, and some species only germinate at certain seasons of the year. The species can be divided into three classes:

(1) Those with no dormancy period. These include chickweed and the usual little ephemerals, and the shed seed of cereals. With these practically all the seeds in the surface inches grow.

(2) Those with a short dormancy period, but which will germinate in autumn; cleavers, speedwell, wild oat, slender foxtail, poppy and charlock are included in this class. Many of these may be induced to grow if the land can be left unploughed until late November; a light harrowing in October will be

helpful, as September rain may have beaten the surface down after the broadsharing.

(3) Those showing a definite dormancy period and which germinate in winter or spring. Many weeds are unfortunately included in this class, and for them stubble cleaning is useless; examples are bartsia, knotgrass, black bindweed and fathen.

It is best to restrict the meaning of the term stubble cleaning to cover only the shallow breaking of the land so that the seeds of annual weeds may germinate, but in practice it is a term which is used very loosely to denote any cleaning operations carried out after harvest. At least three other cases should therefore be mentioned here, though stubble cleaning is a misnomer when applied to them.

Heavy land may, with the frequent growing of autumn-sown crops, become very foul with slender foxtail, corn buttercup, shepherd's needle and wild oat. The case has been mentioned in Chapter I where it was seen that one method of control is to prepare a rough seedbed in autumn, but to defer sowing a crop till spring. The operations would be somewhat as follows: plough as soon as possible after harvest, cultivate in October, drag harrow in early and in late November, and, if possible, plough again shortly after Christmas. In this way successive crops of the weeds may be grown and killed, and a heavily infested field may be almost completely cleared of slender foxtail, which will germinate throughout the autumn months; corn buttercup and shepherd's needle are rather later, and their elimination necessitates working after Christmas, which may, of course, be impossible on heavy land. Some wild oats may also be persuaded to grow by this series of cultivations but it is too much to hope to clear a field of wild oats by any one attack, since their seeds may lie dormant for years.

Where an early harvest has been gathered and dry weather continues subsequently, much cleaning can be done by moving the land several times in September. The field should be ploughed or cultivated as soon as possible, and should be cultivated or drag harrowed at least twice subsequently. In dry weather perennial weeds may be killed very satisfactorily by this procedure, which is a sort of half-fallow, though it would often be described as stubble cleaning.

The next section will be devoted to a method of working perennial weeds free of the soil and burning them. On heavy land this can only be done when the land is uncommonly dry, and is therefore usually only attempted after harvest in a dry season. This again is often referred to as stubble cleaning, though it is better to reserve for it the term 'autumn cleaning.'

WORKING PERENNIALS OUT OF THE LAND AND BURNING THEM

Two perennial weeds very common on arable land are couch and watergrass, and one method of dealing with them is the drastic one of dragging them clear of the soil and collecting and burning them. In this there is one danger to avoid with the greatest care; that is the danger of breaking or cutting up the rhizomes of couch, because each resultant piece may grow, and the last state may consequently be much worse than the first. If rhyzomes are cut at all they should be cut and moved repeatedly so that the resultant pieces cannot get established. Thus trouble will certainly ensue if land foul with couch is disk harrowed (or rotovated) once or twice, but if it is disked five or six times the couch may be almost entirely eliminated.

The procedure for working perennials out of the ground is somewhat as follows:

(*a*) *Separating a layer of the top soil.* This is done with the plough, which should not run deeper than 4–5 in., so that rubbish may not be put too far down. The optimum depth is just sufficient to avoid cutting the couch rhizomes, no severed pieces of which should be seen in the bottom of the furrow.

(*b*) *Breaking the furrow slice.* A cultivator drawn across the furrow will accomplish this, or, on lighter land, a drag harrow may be sufficient. If the land is dry enough, but not too dry, a roll may be used, though it rarely figures in this work.

(*c*) *Separating the rubbish.* The weeds are shaken free of the soil and brought to the surface; possibly a cultivator may be used, but in general a drag harrow with its larger number of tines is preferable.

(*d*) *Collecting the rubbish.* A light harrow is sufficient for this, the harrow being lifted to deposit the weeds where convenient.

This is one, possibly the only one, task which a plain chain harrow will accomplish satisfactorily.

(*e*) *Burning or carting*. In dry weather the rubbish is fairly easily burnt, and the usual method is to have a line of small fires along one or both headlands. Alternatively, it may be collected by the harrow and carted to one or two large heaps, which, once lighted, may burn for a week or more.

(*f*) *The operation may be repeated*. That is to say, a cultivator may be put through the ground once or twice more and another collection and burning made. This does not add very much to the total cost and may just 'complete the job'.

The operation is carried out at various times of the year, but it will be appreciated that on land containing any appreciable amount of clay the weeds cannot be shaken free of the soil, without breaking, unless the soil is dry. Indeed on heavy clay land it is only at the end of a dry summer that success can be achieved, and in such a season the alert farmer takes the opportunity immediately after harvest, and thus probably saves himself from the necessity of fallowing the field; it is the dry summers, when the drought is continued well into September, which provide the opportunity, and then 'twitch' fires are frequently seen. Often, unfortunately, the term stubble cleaning is used to denote this process conducted at that time, but because of the obvious confusion thereby occasioned it is best to describe it as autumn cleaning.

On light land the rubbish can be shaken out in most seasons and the operation—termed spring cleaning—is most commonly performed in April, before a root crop is sown. The initial ploughing is given during the winter, and when the cleaning itself is started it is desirable to push on rapidly with the work, or much moisture will be lost from the soil. To avoid this, and to complete the operation before farmyard manure is applied, it is better to carry out the work in autumn, but that is often precluded by pressure of other work. Sometimes the same methods are used in fallowing, the perennial weeds being dragged out and burnt during the summer; this has little to recommend it, since it suffers from the main disadvantage of a fallow, that a crop is lost, whereas, done in autumn or spring, it will not necessitate the field being uncropped for a year.

It will be realised that often a field is only foul with perennial weeds in parts, and then the procedure is restricted to the portions where it is needed; if the weeds only occur in patches, forking them out by hand may be the best method of dealing with them.

FALLOWING

Light land is sometimes fallowed, the method being to move it repeatedly, and, if foul with perennials, to incorporate the cleaning method just described. But cleaning in autumn or spring and the growing of cleaning crops should enable a light-land farmer to master his weeds without resort to a fallow, which means paying rent and cultivating without any return. On the other hand, a heavy-land farmer cannot grow a large proportion of root crops, and can rarely drag out perennial weeds; on heavy land, therefore, a fallow shift in the rotation may be necessary, and the further discussion of fallows must be understood to refer to heavy land. A variety of perennial weeds may infest the field, but the commonest offenders are couch, watergrass, docks and thistles.

The principle of the fallow is to kill weeds in the clod; that is, to get the land up in a rough cloddy condition in summer, so that in hot dry weather the clods may be dried right out, and then the perennial weeds, rooted in the clods, will be desiccated and will die. The expression 'winter fallow' is therefore a contradiction in terms, and it is very rare that a fallow can be made in spring; it is the months of June to August that must be relied on for the baking weather, though a farmer may be fortunate enough to get the right weather as early as May or as late as September. There are two sorts of fallow, the bare and the bastard fallow, and it is proposed to deal first with the bare fallow. Practically always a bare fallow follows a cereal crop and generally it is followed by an autumn sown cereal.

The procedure in a *bare fallow* is as follows:

(1) The field is left alone from harvest till the following April. If it is ploughed the winter frosts will pulverise the soil, and then the rough clods will be unattainable.

(2) The field is ploughed in April. It is better to invert the soil at the outset. The ploughing should be done when the land

is still wet, so that it may come up in long unbroken slices. The full ordinary depth should be the aim.

(3) A month or so later, when the furrow slice is dried right through, the field is cross-ploughed. In the past cable sets, with a steam engine on each headland, have been employed with very good effect on fallows. These sets have the advantage of great power, since the engine has not to move itself, and they can pull a heavy, deep-working implement across the field without any crushing down of the soil; steam cultivators can work deeper than the bottom of the furrow slice and it has been common when working fallows to 'let one leg down', that is, to set one tine of the cultivator to penetrate to 15 in. or so and burst up the subsoil. Conversion of steam engines to diesel power eliminates the toilsome carting of coal and water but the set still requires a minimum of three men to operate it and the advent of the high-powered, track-laying tractor, capable of pulling the heaviest implement and requiring only one man, has sounded the death-knell of cable work. This is regrettable, in regard to fallow-working, because, though the heavy tractor can pull the implement required, it has to traverse the field itself and in doing this it breaks down some of the clods whose presence is essential at this stage of a fallow. Many farmers prefer the plough because it undercuts all the weeds, but a heavy cultivator will certainly leave the field rougher. The object is to get the field so rough at this stage that it is a misery to walk across it.

(4) The clods are stirred once or twice in hot dry weather, so that they may become dried right out. This stirring is done with the cultivator or long-tined harrows (occasionally by the plough), and the operations should be timed carefully to get the maximum drying effect; often, however, the crucial opportunity is offered during haytime or when the tractor or horses cannot be spared.

(5) The field is ploughed again just before, or just after, the corn harvest. Despite the rough condition in late May, the alternate wetting and drying, and heating and cooling which they receive, together with such soil cultivation as is done, will work the clods down, and it will generally be 'tilthy' by late July. It is to be hoped that the perennial weeds will have been

killed earlier, but in the mould that is produced annual weeds will appear in a showery time, and one ploughing will account for them. It is better if this can be done after, rather than before harvest, both to kill the maximum number of annuals and to bring up some clod to the surface, because after a fallow a field is likely to have too fine a tilth for autumn sowing.

The effectiveness of a fallow is entirely dependent on the weather. In a wet summer the whole thing may be a waste of time, and may make the field fouler than ever, but in a dry summer couch and watergrass can be completely eliminated. In making a bare fallow it is important not to lose the clod too early in the summer so that a long period is available for wilting the perennials; this point is further important because of the wheat bulb-fly which selects fairly smooth bare ground for laying its eggs about the month of August.

The *bastard fallow* differs from the bare fallow in that it occupies only the second half of the summer. It may be taken after a seeds ley, after a silage crop or after oats and tares cut for hay. The procedure is to burst up the land as soon as the crop is off, using a plough if the land is not too dry, or heavy cultivator in a dry summer. If the plough is used, deep cultivation, by cable or tractor, should follow about a fortnight later, and then the field is in the rough cloddy condition required. As opportunity presents itself, and the weather indicates, the clods should be moved once or twice during August and early September. No new principle is involved, but it will be appreciated that the whole procedure is telescoped as compared to a bare fallow. The effectiveness of the bastard fallow may be high, but it depends entirely on a dry latter half of summer. July and August are often hot, dry months but September rarely is, so that the rapid breaking of stubbles after harvest, referred to earlier, is rarely comparable to a fallow in its cleaning effect.

The following are the advantages of a fallow:

(*a*) *The cleaning effect*, which is the main reason for a fallow in this country.

(*b*) *The saving of moisture*. Bare ground loses less moisture through evaporation than cropped ground loses through transpiration, and this is the main reason for fallowing in dry farming (p. 20). The moisture saved would be unimportant in this country.

(*c*) *The rendering of plant food available.* This effect is an important one, mineral plant foods being rendered available by the action of the weather on the cultivated soil, and nitrogen by the action of bacteria when the soil is warm; the nitrogen, however, is liable to be washed out when the autumn is wet, and this is one of the reasons for growing mustard after a fallow, though leaching is less likely to occur on clay land. The effect is well illustrated by the yields on Broadbalk Field at Rothamsted; the unmanured plot, having grown wheat for 77 years, had fallen in yield to 12 bushels per acre, but after fallowing the yield rose in the following year to 28 bushels per acre. The effect was, however, only transitory, because in the next year the yield fell right back to its previous level.

(*d*) *The field is ready for autumn sowing at the end of the summer.* This is more important than appears at first sight, because on clay land the farmer is apt to get behind with his work; late drilling accentuates the normal lateness of harvest and so on in an ever-increasing lag, and to have at least one field ready for sowing early is no mean advantage.

(*e*) *The improvement of soil texture.* At the end of a fallow the soil should be in a good 'crumby' condition, and this ameliorative effect may, in the absence of any soaking wet period, persist for a long time.

The following may be reckoned as the disadvantages of a fallow:

(*a*) *The cost*, and lack of immediate return.

(*b*) *The tilth is often rather too fine for autumn corn*, and the soil may run together if a wet winter follows.

(*c*) *The wheat bulb-fly is encouraged.* This pest lays its eggs on bare ground which has a fairly fine tilth from mid-July to mid-September, and the grubs which hatch out eat the central shoots of wheat from the following February till April. Great damage is sometimes suffered from this pest in wheat following a fallow, and one of the objects of taking a catch crop of mustard at the end of a fallow is to cover the ground in August; though common on light land this procedure is rarely followed on heavy clay.

Wheat is the usual crop to follow a fallow despite the pest just mentioned. Winter oats are sometimes grown, but their value per acre is lower than that of wheat, and the farmer naturally looks for some reasonable return for his fallow in the succeeding crop. A case can scarcely be made for a spring-sown crop, because of the loss of nitrogen which might occur by leaching before the crop was drilled, and also because after its summer working clay land might lose its good texture during a wet winter, and spring sowing would be impossible. These two risks are too great to run, though against them might be set the possibility of killing slender foxtail, corn buttercup, shepherd's needle and wild oat, none of which will suffer in the fallow, but any of which may germinate in serious numbers in the following autumn. Beans suggests itself as a suitable crop, as being immune to the wheat bulb-fly, and as providing a chance for spring hoeing if the above autumn- and winter-germinating annuals appear; but beans, again, is a crop of lower value than wheat, and, being leguminous, would not give the full return for the nitrogen rendered available in the fallow.

Comparing the bare and the bastard fallow, the latter has the advantage in not wasting a whole year, and as not leading to such a fine tilth at its conclusion; it is also less liable to lead to trouble from wheat bulb-fly, as the pest appears to prefer smooth ground, and in a bastard fallow the land is still very rough when it lays its eggs. On the other hand, the bare fallow gives the better chance for killing the perennials, because a dry time at any part of the summer will suffice, and will also lead to a greater provision of available plant food.

SPRAYING

Spraying is a powerful weapon to which the farmer can resort in certain cases and it is one which has been developed remarkably in very recent years. Spraying is selective weed killing, that is, the application of a chemical to a weed-infested crop with the object of killing the weeds without harming the crop. This idea is obviously attractive and is by no means new, but recently very great advance has been made.

The weed against which early efforts with spraying were directed was charlock and that still remains one of the chief targets. Charlock seeds are oily and retain their viability in the soil for very many years, seedlings appearing in vast numbers when suitable conditions arise. Germination can occur at practically any time when the surface of the soil is moved, but if the charlock plants appear in autumn they need occasion no alarm, because the winter frosts normally kill them. It is when they appear in spring that some action is imperative, because they are often so thick on the ground that the crop is utterly ruined. In root crops the repeated hoeing is a fairly effective method of control, but in corn crops a bad infestation cannot be controlled in this way, because so many of the weeds are growing in the corn rows. It is usually spring-sown corn which suffers most, though when autumn corn is harrowed in March a very thick growth of charlock occasionally appears. Some farmers make a practice of harrowing spring-sown barley just before it appears above the ground and undoubtedly this is very effective in killing charlock seedlings; the practice, known as charlock harrowing, is a drastic one, however, and carries with it grave risk that the young barley seedlings may also be killed. Spraying is the only practicable means of attack in a bad case. Fortunately, the broad, rough charlock leaves catch and hold much more of the spray than the narrow waxy cereal leaves, so that the latter rapidly recover whilst, under favourable conditions, the charlock dies. The powder form of calcium cyanamide has been used with some success; application is best made on a dewy morning and for full effect this must be succeeded by a sunny day. For many years copper sulphate (known to farmers as bluestone) was widely used for spraying against charlock. From 50 to 100 gallons were applied per acre, the strength of the solution being from 3 to 5 per cent. For good results with copper sulphate the charlock plants must be young, possessing not more than four leaves, and the spraying must be carried out on a fine day; rain within a few hours of spraying will ruin its effectiveness.

With the chemicals recently introduced a wide range of weeds can be attacked and spraying is not now limited to the control of charlock; nor is it now essential to spray when the weeds are very young, though that remains very desirable both because they are easier to kill then and also because the longer they are allowed to grow the greater damage will the crop suffer from their competition.

It was demonstrated as early as 1911 that sulphuric acid was effective against a number of weeds and that it could be used without killing a corn crop. The action is much quicker than that of copper sulphate and a few hours of sunshine after application suffices to ensure a good kill. The main disadvantages of sulphuric acid are its corrosive action on the machine and the difficulty and danger of handling carboys of strong acid on the farm. Furthermore, there appears to be considerable damage to the corn crop, though recovery is generally rapid and experiments have shown that cereal plants that have been sprayed with sulphuric acid attain their full yield. Where the crop was foul with weeds and the acid has eliminated them great increases in yield have resulted. Sulphuric-acid spraying did not become common in this country until about 1930, when some contractors equipped themselves with lead-lined sprayers and undertook the work. The strength of acid used should vary with the weather and according to the weed species most prevalent; in regard to the latter the reader is referred to Table IV.

Copper chloride has proved preferable to copper sulphate and has been used to a slight extent at concentrations of from 1 to 4 per cent; higher concentrations than 4 per cent injure cereal crops. The action is quick and the solution is harmless to skin and clothing though it is very corrosive to machines, which must be thoroughly washed out after copper chloride has been used in them.

A big step forward was taken in spraying technique with the introduction of dinitro-ortho-cresol compounds (commonly referred to as DNOC). These have been used as yellow dyestuffs for a long time and were known to be insecticides but it was not until 1932 that their value as selective weed killers was demonstrated by French workers; their use for spraying was only developed in this country during the later years of the War of 1939–45. DNOC compounds are sold (as proprietary weed killers) in the form of thick pastes and are non-corrosive, but they produce bright yellow stains which are temporary on the human body but permanent on clothes. These compounds can have most serious effects on men working with them continuously and deaths have occurred. It is most important that the precautions laid down should be most stringently followed; generally

TABLE IV. RELATIVE SUSCEPTIBILITY TO SELECTIVE WEED-KILLERS OF WEEDS COMMON IN CEREALS

Common name	MCPA	2·4-D	DNC	Dinoseb	Sulphuric acid
Stinking mayweed	MR	MR (E)	MR	MR	MR*
Orache	MS	MS	MS	MS	MS*
Wild oats	R	R	R	R	R
Field cabbage	MS	MS	—	—	—
Shepherd's purse	S	S	S	S	S*
Fathen	MS	MS	MS	MS	MS*
Corn marigold	R	R	MS	MR	MR*
Wild carrot	MR	MR	—	—	—
Treacle mustard	VS	VS	S	S	S
Fumitory	MS	MS	S*	S	R
Hempnettle	MS	MR	S	S	S
Cleavers	R	R	MS	MS	S
Scentless mayweed	MR	MR (E)	MS	MR	MR*
Poppy	MS	MS	S	MS	R
Knotgrass	MR	MR(MSE°)	MR	MR	MS*
Black bindweed	MS†	MS†	MS	MS	S
Redshank	MS†	MS†	MR	MR	MS*
Corn buttercup	S	S	MR	MR	MR
Wild radish	S	S	S	MS	MS
Shepherd's needle	MS	MS	MS	MS	MR
White mustard	VS	VS	VS	VS	—
Yellow charlock	VS	VS	VS	VS	VS
Sowthistle	MS	MS	MR	MR	MS*
Spurrey	MS	MS	MS	MR	S
Chickweed	MR	MR	MS	MS	MS
Pennycress	VS	VS	S	S	VS
Annual nettle	MR	MR	S	MR	S*
Speedwell	MR	MR	S	MS	S
Heart's ease	MS	MR (E)	MR	MR	—

VS = very susceptible. S = susceptible. MS = moderately susceptible
MR = moderately resistant. R = resistant.
(E) indicates that ester is more effective than amine.
MS† when *young* seedlings only otherwise MR.
MSE° = MS if ester formulation is used.
* A wetting agent should be included in the spray solution for these weeds.
Reproduced from the Ministry of Agriculture's Advisory Leaflet 436 by permission of the Controller of H.M. Stationery Office.

DNOC is only applied by contractors who have protective clothing and protective cabins on their tractors. The sodium salt has been widely used but it is slow-acting and relatively ineffective at temperatures below 50° F. The ammonium salt is more effective at lower temperatures and the sodium one can be 'activated' by adding ammonium sulphate to the solution; the solution required per acre is 8 lb. of sodium DNOC and 10 lb. of ammonium sulphate in 100 gallons of water.

A most interesting and promising line of advance was opened up in 1940, when it was found that certain growth-promoting substances, known to encourage quick rooting of cuttings, killed some weeds and did not damage cereals. Intensive research narrowed the field down to derivatives of phenoxyacetic acid and at the present time two of these derivatives are much used, being put on the market as proprietary weed killers. The two compounds are known to chemists as 2-methyl-4-chloro-phenoxyacetic acid and 2:4:di-chloro-phenoxyacetic acid; these are not the sort of names that are bandied about among farmers and the compounds are referred to as MCPA and 2·4–D respectively, and by the trade names of the proprietary weed killers containing them. The action of these compounds is very different from that of all other herbicides. The weeds are distorted and killed very slowly so that effectiveness is independent of the weather, whilst the cereal crop is normally completely unaffected; these compounds are absorbed through the roots of plants and hence will kill minute seedlings that have not yet come above ground. An important point is that extremely small amounts of MCPA and 2·4–D are required, not more than 2 lb. per acre. The material has been applied as a dust, 1 lb. of the chemical being 'diluted' with about 1 cwt. of inert material by the manufacturer. These dusts can be applied with an ordinary manure distributor but they are not so efficient in action as solutions, in which form herbicides are now normally applied. Hormone weed killers are not toxic to man or animals and are not corrosive. Very occasionally the cereal crop may suffer. If sprayed too early there may be a proportion of malformed ears at harvest and if too late there may also be a diminution in yield. Wheat and barley should have at least two expanded leaves on the main shoot before they are sprayed but oats can be sprayed with MCPA when they have one full leaf; all cereals are liable to damage if they are sprayed after they have started to shoot. There is more danger of harming the crop if cereals are sprayed within 10 days, before or after, of cultivation such as rolling or harrowing.

A frequent case which presents difficulty is that where a cereal crop is undersown with clover and the field is weed infested. The sprays mentioned above will kill legumes but they have been successfully used on undersown crops. If the undersowing is to be done when the cereal has become well established the field may be sprayed with sulphuric acid or copper chloride immediately before the small seeds are sown, but there must be an interval of at least 10 days between spraying with DNOC and sowing clover, whilst with hormone herbicides the interval must be 6 weeks, which can rarely be allowed in practice. There are cases where the infestation with charlock is so dense that young clover plants are sheltered from the spray, practically all of which is caught and held on the charlock leaves; if the young clover plants are well established they may survive but there is great risk of losing them. There are two newer forms of herbicide which will meet this case. One is dinitro-secondary-butyl-phenol (DNBP or dinoseb) which does not kill legumes and has proved useful for undersown cereal crops, for mixed crops of cereals and pulses and for lucerne crops; it is, however, expensive and men using it must be protected. The other form arose from the important discovery that phenoxybutyric derivatives could be used as herbicides. These have no lethal effects themselves but some plants have enzymes which reduce these compounds to the corresponding phenoxyacetic ones, which then exert their effect and the plants die. Clovers do not possess these enzymes and so are immune, but the same is also true of certain weed species so that the effective range of phenoxybutyric compounds is not so wide as that of phenoxyacetic ones. There appears, however, to be a big field of usefulness for these new herbicides which are known as MCPB and 2·4 DB. They, or similar ones still to be developed, may prove valuable for controlling weeds in grass fields; spraying to kill weeds in pasture is done fairly frequently but with mixed success, there always being the danger of killing out the clovers.

In general, the herbicides mentioned so far are effective against many dicotyledonous plants but do not affect monocotyledons; that is why they have proved so useful for cereal crops. Others are appearing which act in the reverse way, that is, they are more deadly to monocotyledons than to dicotyledons; these hold out hopes of efficient spraying for root crops but it cannot be said that this further advance has yet been achieved. The method is what is called pre-emergent spraying with chemicals such as isopropyl-phenyl-carbamate (IPC), its chlorinated derivative (CIPC) and trichloroacetic acid (TCA). These are applied to the land either before the seed is sown or between sowing and the emergence of the crop plants and with careful timing they may succeed in giving the crop a clean start. Although pre-emergent spraying is already practised by some progressive farmers the method cannot yet be said to be out of the experimental phase. Trichloroacetic acid is being used to control couch but only where the field is bare of crop and not to be sown for 2 or 3 months; two applications, with cultivation in between, are necessary and the control is moderately good but very costly.

The exact area of crops which is sprayed every year is unknown but it is estimated at around 2 million acres in Great Britain, which indicates the great impact the new herbicides have had on farming (particularly MCPA which is by far the commonest chemical used). A practice which has spread so widely in little over 10 years must obviously have proved itself of great value and every year new chemicals and new methods are described so that it is fairly safe to prophesy that it will soon be possible to kill any weed by spraying and even that that may be done in any crop. There are those who think that weed control in the future will be entirely a matter of using the right chemical at the right time and that the old established methods discussed earlier in this chapter will soon have nothing but historical interest. This is going much too far. The traditional methods of keeping land clean—rotations and cultivations—have one very strong argument in their favour, namely that they are desirable for other good reasons; their cleaning effect is almost a by-product involving little, if any, additional expense. Spraying is costly, the chemicals used varying in price from about 10*s*. an acre sprayed up to £7 or £8; application may not cost much to the farmer equipped with a sprayer. But for efficient work the farmer has to know what chemical to select, what formulation of it is best, how much to use and when to use it. A farmer with many other things on his mind has no possible chance of keeping himself up to date in chemical weed control; indeed, it is doubtful if anyone can claim to be abreast of all developments except those who attend the British Weed Control Council biennial conferences and they only for a few weeks following one of the conferences. The country is fairly well covered by spraying contractors who take pains to keep in some touch with developments and many farmers come to rely on these men not only to do the job but also to advise on the chemical to use. Contractors naturally allow themselves in their charges a good margin over cost and if they are to supply the knowledge as well they have a clear right to charge for that too. Contract rates for spraying are a frequent cause for grumble among farmers but it would be optimistic to expect them to decline to any considerable effect.

The position may get simpler as knowledge advances but it would be unrealistic to hope for one chemical that will kill all weeds and not harm any crop. These chemicals are selective and the constant use of any one of them may eliminate from a field the weeds to which it is deadly, only to provide opportunity for others to develop in profusion. The present serious position in some parts of the country in regard to the wild oat is evidence of this; again, chickweed was an insignificant pest in the past but has become a menace in many districts because it is resistant to hormone herbicides. If chemical weed control becomes the main method it is certain that there will have to be rotations of herbicides and these will call for yet more knowledge. Concern has been felt that the application of these little known chemicals to the land may have deleterious effects on fertility. There appears to be no cause for alarm over this since soil organisms quickly break down the chemicals to innocuous substances; this is not to say, of course, that there is no possibility of a chemical being found which may have a lasting and harmful effect on the land.

The expressions 'high volume' and 'low volume' are frequently heard in regard to spraying. With the former the appropriate amount of chemical is applied to the land in 80–100 gallons of water per acre, while with the latter the amount of water may be as low as 5 gallons per acre. High volume spraying clearly gives a better chance for even distribution of the chemical over the field and in some cases is demonstrably the more effective method but it suffers in that it involves much carting of water and that the sprayers for it are usually expensive machines. Low volume sprayers are cheaper in cost and in use and economists have calculated that an average annual demand of 30 acres spraying will justify a farmer in purchasing one; to justify buying a sprayer which can be used for high or low volume work 66 acres are necessary. The introduction of low volume spraying has accentuated one serious danger in the use of herbicides, that of spray drift. This has always to be reckoned with, particularly in low volume work, for such sprayers produce very small droplets and these may drift in a light breeze for considerable distances and harm susceptible crops of the same farmer or, worse still, of his neighbours. It is easy to recommend that spraying should only be done on still days but the weather may be windy for a week or two covering the optimum time of application; even on still days there may be vertical air currents that will carry the droplets upward for them to descend again a mile or more away from the spraying site. Tomato growers have suffered most from spray drift and there have been cases where heavy damages have been awarded to them; how frequently harm from spray drift occurs is unknown but it must be quite common. The injury only becomes evident two or three weeks after the spraying has been done and it may have serious effect on a crop and yet be difficult to identify as spray damage; even then it is hard to prove that a particular farmer was responsible when several may have been spraying at the same time. The avoidance of spray drift is obviously important. All that can be done is to choose calm days for spraying, to have spray booms as near to the ground as possible and, where practicable, to fit shields to the booms; even then a farmer who sprays is wise to take out a third party insurance policy.

Chemical weed control has come in the nick of time, just as rising labour costs have made hand weeding almost prohibitive; its great value cannot be disputed, yet it is important to realise that it is only one weapon, admittedly a very powerful one, in the fight against weeds. All that has been attempted above is to give a general account of a new and very important development. Table IV shows for a short list of common weeds the effectiveness of the more common herbicides available at present.

A full treatment of the subject would be very lengthy and within a few months would become seriously out of date. Fortunately the British Weed Control Council issues a Weed Control Handbook annually; this is a complete and authoritative production covering the whole field and for the latest information reference should be made to the most recent issue.

PLANT POISONS

These are classed separately from the above because the object is to kill all the plants on an area. A variety of chemicals can be used. Arsenic is commonly applied to garden paths, but it is an expensive and a dangerous material to have on a farm. The most useful compounds are the chlorates of sodium, potassium and calcium. Solutions of these suffer from the disadvantage that cotton soaked with them is explosive when dried; thus overalls should be worn when applying these solutions and should afterwards be washed. There is some difficulty in getting solutions of the desired strength—2–6 per cent—and with potassium chlorate, in particular, hot water is desirable in making the solution. Fortunately the chlorates are harmless to men and to stock, but for plants they are potent poisons. The normal movement of water in a plant is in an upward direction so that, unless the plant is wilting, solution absorbed by the leaves will not kill the root-stocks of perennials. Where an infested area is sprayed, however, much solution will soak into the ground and kill the root-stocks which absorb it; the amount and concentration of solution required to be effective will be determined by the permeability of the soil, by rainfall and by the position of the roots.

It has been said that the chlorates kill all the plants on the area sprayed; this, unhappily, is not quite true, because the wild onion, one of the worst scourges of some areas, though looking sickly after the application, eventually recovers. If an area is sprayed with a chlorate it will take some time for the poison to become eliminated; the length of time clearly depends on the rainfall, but it would be unwise to sow a crop on the area before three months, at least, have elapsed. This must also depend on the amount applied; a case has been reported in which nettles growing in a crop of Italian rye grass were sprayed successfully with a weak solution (1 per cent) of sodium chlorate, with little damage to the rye grass.

The chlorates are rather expensive, so that to apply them to a whole field would rarely be economic. They have a rightful use, however, for patch-work application, and the commonest case for which they are advised is that of nettles growing in clumps on a grass field. Often these clumps are so thick that little spray will fall on the ground, and probably a further application will be necessary to eradicate the clumps. Another good opportunity for the use of chlorates is presented by hedgerows, which often carry a variety of luxuriantly growing weeds; one man can spray a great length of hedgerow in a day, but care must be exercised to avoid spraying the hedge itself, which may also be killed. Thorn often appears on pastures and is very difficult to eliminate by digging, because of the depth of its roots. One man can deal with it at great speed if he is armed with a watering can, and squirts a little chlorate solution over a small circle of soil around each plant.

As with the selective herbicides so also with plant poisons new chemicals are coming into use. Borates, thiocyanates and sulphamates have been used but at present the most promising one appears to be para-chloro-benzene-dimethyl-urea (CMU). This will kill nearly all plants and though it is slow in action it is extraordinarily persistent in the soil. This plant poison has not been used much as yet and would seem to be more suitable for gardens rather than for farm land. It must not be used within the range of a tree's roots since it can be very deadly to trees.

THE COMMON WEEDS AND METHODS FOR THEIR CONTROL

Many weed species plague the British farmer, but it is not proposed here to deal exhaustively with them, with their botanical characteristics, nor the methods by which they may severally be controlled; for a full treatment of the subject, which still requires much elucidation, the reader is referred to books devoted entirely to weeds. But the great importance of cleaning in crop husbandry necessitates some reference to those species most prevalent in the country. The difficulties of this are to decide which weeds to include in a short list, and to arrange the selected ones in any order of importance, because prevalence and the harm wrought must both be considered. The order in the two lists below is admittedly somewhat arbitrary, but it represents an attempt to assess prevalence and harmfulness combined, the farmers' worst enemies appearing high in the lists. In justification of the placing it can be urged that no order of demerit can have any objective reality, because there is no common yardstick for comparison. Chickweed, for instance, is almost universal but does not usually do much harm, though it has become very troublesome on some good soils in recent years; hoary pepperwort, on the other hand, though as yet confined to fairly small areas, spreads at an amazing rate and is difficult to eradicate. There are several distinct species of some of the weeds included, but they are not listed separately, except when the methods of control differ.

In order to deal with the matter summarily, a form of shorthand will be employed, using letters to denote the various cleaning methods the farmer can adopt. Selective herbicides and plant poisons will not be included since there are now, or shortly will be, chemicals available for dealing with practically all weeds. The methods have already been described, and for arable land the following are available:

(A) The use of clean seed.
(B) Stubble cleaning.
(C) Working the land in spring, before sowing a root crop.
(D) Growing root crops and hoeing them thoroughly.
(E) Hoeing corn and pulse crops.
(F) Preparing a rough autumn seedbed, but not sowing a crop until spring.
(G) Fallowing.
(H) Dragging out and burning.
(J) Pulling from corn crops (and, occasionally, from root crops in late summer).
(K) Putting the field down to a long ley.
(L) Liming.
(M) Draining.
(N) Working the land in February and March.
(O) Digging out by hand (where the infestation is restricted to small areas).

Effectiveness will depend on the possibility of carrying out the treatment in good conditions, so that the various methods are listed below in alphabetical order for each species; a method which may be slightly, but usually is not very, effective, is denoted by its letter enclosed in brackets.

Couch (*Agropyrum repens* Beauv.)	C, G, H, (K)
Watergrass (*Agrostis stolonifera* L. var. *alba*)	C, G, H
Charlock (*Sinapis arvensis* L.)	B, C, D
Creeping thistle (*Cirsium arvense* Scop.)	B, C, D, E, G
Wild oat (*Avena fatua* L.)	A, B, E, F, K
Dock (*Rumex obtusifolius* L.)	A, B, C, D, G, J
Poppy (*Papaver* spp.)	A, B, C, D, E
Slender foxtail (or black grass) (*Alopecurus agrestis* L.)	A, B, E, F, K
Cleavers (or goose grass) (*Galium aparine* L.)	B, C, D, E, K
Chickweed (*Stellaria media* L.)	B, C, D, E, K
Corn buttercup (*Ranunculus arvensis* L.)	A, E, F, K
Fathen (*Chenopodium album* L.)	C, D, J, K
Runch (*Raphanus raphanistrum* L.)	B, C, D
Wild onion (*Allium vineale* L.)	A, (E), M, O
Onion couch (*Arrhenatherum avenaceum* var. *bulbosum* L.)	(A), G, H, O
Coltsfoot (*Tussilago farfara* L.)	C, D, E, M
Shepherd's needle (*Scandix pecten-veneris* L.)	A, E, F, K
Campions (*Lychnis* spp.)	A, C, D, E, J
Bindweeds (*Convolvulus* spp.)	C, D, E, K
Mayweeds (*Anthemis* spp. and *Matricaria* spp.)	B, C, D, E, K
Knotgrass (*Polygonum aviculare* L.)	C, D, E, K
Speedwells (*Veronica* spp.)	B, C, D, E, K
Corn gromwell (*Lithospermum arvense* L.)	C, D, E, K
Sorrel (*Rumex acetosa* L.)	(B), (C), (D), L
Spurrey (*Spergula arvensis* L.)	(B), (C), (D), L
Hoary pepperwort (*Lepidium draba* L.)	(C), (D), (E), O
Shepherd's purse (*Capsella bursa-pastoris* D.C.)	B, C, D, E, K
Groundsel (*Senecia vulgaris* L.)	B, C, D, E, K
Hemp nettle (*Galeopsis tetrahit* L.)	C, D, E, K
Willow weed (or redshank) (*Polygonum persicaria* L.)	C, D, E, J
Fumitory (*Fumaria officinalis* L.)	B, C, D, E
Horsetail (*Equisetum* spp.)	E, M
Dodder (*Cuscuta* spp.)	A, O

Corn marigold (*Chrysanthemum segetum* L.)	A, B, C, D, E, J
Sow thistle (*Sonchus oleraceus* L.)	C, D, E, J
Dead nettle (*Lamium* spp.)	C, D, K
Cornflower (*Centaurea cyanus* L.)	C, D, E, K

In some cases no one attack can do much good and a whole campaign must be undertaken and faithfully sustained to bring a particular weed under control. Wild oats are very troublesome because of the longevity of their seeds in the soil and the failure of a large proportion of them to germinate when the conditions are made suitable for their growth; too often it is only when cleaning operations are over and a crop planted that the laggards come up, to mingle indistinguishably with corn and populate the field afresh. Two successive root crops, with well-worked seedbeds, and with proper hoeing, will reduce the population considerably but will not often eradicate them, so that without constant watchfulness they become as bad as ever again; in some areas especially prone to wild oats they remain a serious plague despite high and intelligent farming. Wild onions are very pernicious. On light land or good loam they refuse to grow and are never seen, but on clay soil they are a real menace. They reproduce themselves by aerial bulbils carried on stalks 2 or 3 ft. high, and also by offsets formed at the foot of the stalk. Many a farmer has expended much labour by pulling them in June or July from a corn crop, the whole effort being wasted since the offsets are left in the ground to give a full infestation of the field in the following year. They can be completely eradicated by avoiding ploughing in September and October and growing a succession of six spring-sown crops; but it is difficult, often impossible, to act on this recommendation because the heavy land they infest is unsuitable for spring sowing. They survive in a grass field and for a short period in spring are eaten with relish by cows, whose milk they taint. A successful ley of cocksfoot and lucerne, cut at least twice a year for 4 years may exterminate them, the times of cutting such a ley (late May and July) being apparently vulnerable periods in their growth. Formation of aerial bulbils is prevented and the offsets do not harden and hence have no dormancy period; the power of the offsets to remain dormant in the ground, to grow out at various times, is one of the main reasons why wild onions are so difficult to eliminate. For onion couch the recognised control is to fallow the field for two years running, but this requires dry years for complete success; in any case this treatment is very expensive and hence uncommon with the result that the weed remains rampant in some areas. A method that has been used with success is to ridge up the field in autumn and to throw out the knotted rhizomes in spring with a potato spinner, and then to pick them up and cart them off the field.

On grassland the methods which the farmer can adopt are:

(P) Spudding.
(Q) Cutting above ground. In the case of annuals this must be early enough to prevent them from ripening seed. In the case of perennials the time must be chosen so that the root-stocks are exhausted as much as possible, the best time usually being June or July; the treatment is much more effective when it is repeated later in the same year, and it will probably have to be continued for several seasons.
(R) Intensifying management by manuring and by heavy grazing.
(S) Heavy grazing by sheep in early spring.
(T) Application of nitrogenous manure.
(U) Digging or pulling out by hand.
(V) Heavy harrowing in winter.
(W) Use of plant poisons.
(X) Breaking up the field and taking an arable rotation.
(Y) Draining.
(Z) Liming.

Again the list must be taken as only roughly indicating the relative prevalence and harmfulness of the more important species.

Watergrass (*Agrostis stolonifera* L. var. *alba*)	R, V, X, Y, Z
Creeping thistle (*Cirsium arvense* Scop.)	P, Q, X
Stinging nettle (*Urtica dioica* L.)	Q, R, U, W
Buttercups (*Ranunculus* spp.)	Q, T, (Y)
Spear thistle (*Cirsium lanceolatum* Scop.)	P, Q
Ragwort (*Senecio jacobaea* L.)	(Q), S, U
Bracken (*Pteris aquilina* L.)	Q, R, X, Z

Wild onion (*Allium vineale* L.)	(S)
Thorn bushes (*Crataegus monogyna* L.)	U, W
Tussock grass (*Aira caespitosa* L.)	U, Y
Yarrow (*Achillea millefolium* L.)	T
Daisy (*Bellis perennis* L.)	R, (T), X
Plantains (*Plantago* spp.)	T, X
Hawkbit (*Leontodon* spp.)	P, Q, R
Yellow rattle (*Rhinanthus crista-galli* L.)	Q
Wild carrot (*Daucus carota* L.)	P, R
Ox-eye daisy (*Chrysanthemum leucanthemum* L.)	Q, R, X
Sedges (*Carex* spp.)	Y
Red bartsia (*Bartsia odontites* Huds.)	Q, R
Knapweed (*Centaurea nigra* L.)	Q, R
Sorrel (*Rumex acetosa* L.)	Z
Bramble (*Rubus fructicosus* L.)	(Q), U, W, X
Self heal (*Prunella vulgaris* L.)	Q, R
Quaking grass (*Briza media* L.)	R, Z
Rest harrow (*Onorus spinosa* L.)	Q, R
Burdock (*Arctium lappa* L.)	Q, R
Scabious (*Centaurea scabiosa* L.)	R
Cowslip (*Primula veris* L.)	(Y), (Z)
Gorse (*Ulex Europæus* L.)	U, X, (Z)
Broom (*Cytisus scoparius* Link.)	U, X

To the above might be added certain grass species which form a large proportion of the mixture on poor grazing land, where they may, indeed, be of some value, and to which good pasture tends to degenerate; the control of these species is, however, chiefly a matter of well-arranged grazing and adequate manuring. The most satisfactory, and often the quickest, way of eradicating grassland weeds and poor grass species is to plough up the field, put it through an arable rotation and lay it down to grass again; even then it will soon revert to its original condition unless deficiencies in lime, drainage or general management are corrected. Hedgerow weeds such as chervil and cow parsley have been omitted from both lists, as there should be no difficulty in confining them to their natural habitats.

Orders made under the Corn Production Acts (Repeal) Act of 1921 have empowered the Minister of Agriculture (who delegates his functions in this respect to the County Agricultural Committees) to require a farmer to destroy certain injurious weeds; if the farmer does not comply he becomes subject to a substantial fine. The weeds concerned are:

Ragwort,
Spear thistle,
Creeping thistle,
Curled dock,
Broad-leaved dock.

Though several thousand cases are reported to County Agricultural Committees annually, nearly all of them are settled by arrangement, and the prosecutions undertaken are very few; the powers, in fact, are valuable for occasional use in a persuasive manner.

The old proverb 'A stitch in time saves nine' is particularly applicable to keeping land reasonably weed free; neglect of the ordinary methods of control not only leads at once to lowered crop yields, but soon necessitates extraordinary steps which are costly. With some weeds great trouble is saved if the farmer is alert, and attacks them vigorously at their first appearance. Wild onions, for instance, may be inadvertently introduced to a farm (very possibly in seed wheat) and appear in a few isolated patches. Men armed with forks and buckets can, in a short time, dig them out and carry them off the field, and if this is done in February and March for two or three years the infestation may be completely conquered; but if nothing is done for several years the onions will spread all over the field, and then no satisfactory methods of eradication are economically possible, and the farmer must resign himself to selling corn, and even milk, that stinks. Hoary pepperwort is another very vicious enemy, and the most intensive methods, such as digging out by hand, are fully worth while in extirpating the weed from a small patch where it has established itself. The important point with weeds like this is to observe them directly they appear, and to descend on them forthwith in full force.

CHAPTER IV

TILLAGE AND THE REQUIREMENTS FOR A SEEDBED

Though nothing is more fundamental than tillage, it remains the part of crop husbandry which is still more purely an art than any other. Practical experience led to a series of beliefs which it was possible to assemble into a more or less coherent story of the objects and benefits of tillage. But during recent years science has impinged on the subject, and it cannot be said that the position has thereby been clarified; in fact, science has succeeded in undermining some of the practical man's tenets, and has not, as yet, offered him anything in their place. That being the case, it is not proposed to delve deeply into a mass of contradiction, but to deal with the principles of tillage in the briefest possible manner, albeit that they lie at the root of crop husbandry.

If not the chief, one of the most important objects of tillage is the destruction of weeds; it has already been seen that this can be accomplished by the continual moving of the land, but in these days of expensive labour a farmer should choose and time his operations so that he may keep his land clean with few cultivations. The plough will bury, in addition to weeds, crop remains and manure, and the incorporation of these with the soil is another important object of tillage. Thus far there is no room to cavil, but in what follows uncertainty looms largely.

Cultivation loosens the soil, and up to a point this is clearly necessary, so that the seed may be put into the ground and covered up. This only requires a loose condition in the top inch or two, but deeper movement is regarded as important in facilitating the penetration of plant roots into the soil. There is no doubt that soil containing an appreciable amount of clay may set so solidly that root growth is mechanically inhibited, and that, in general, loosening of the soil does produce better developed root systems; although a few roots will penetrate very firm ground—and are, in fact, normally to be found in the undisturbed subsoil—proliferation is more abundant in the tilled layers. It is, of course, impossible to loosen the soil to the full depth to which the roots of a healthy plant should penetrate, but looseness to the normal depth of ploughing will provide at least some encouragement for roots of even the weaker rooting species. It will be necessary to return to this point later, when it will be seen that loosening, like everything else, can be overdone, and that the soil, having been loosened, should be made firm again, so that the plant roots may be in intimate contact with it; it will readily be appreciated that the state of firmness required is very different from the solid condition of unmoved land.

Tillage has an effect on the movement of water in the soil, and it is in this connection, particularly, that soil scientists have shaken the clearly defined older ideas. The movement of water in soils was held to resemble that in fine glass tubes, in which it can easily be demonstrated that water will rise to considerable heights, by what is known as capillarity; the finer the tube the higher will the water rise. The view was that the little spaces between the soil particles formed a series of rather disconnected, but fine, capillary tubes, and that therefore water would rise to the surface from the lower depths. The greater the proportion of clay in the soil the finer would the tubes be, and hence the greater the height through which water would rise; consolidation of the land, by making the pore spaces smaller, would serve to encourage the capillary rise of water. Here was a convincing reason for aiming at consolidation in a seedbed, and rolling was commonly recommended on the score that it would bring up the water. But careful experiments have failed to detect this upward movement of water, except over short distances and through soil so wet that it would be of no use and on which a farmer would not dream of using a roll. There are, of course, soils in which there is a definite water table within a few feet of the surface, but there are many others in which free water is only found at a great depth, which, in fact, may be drier at a depth of a few feet than at the surface. When it is added that the roll only consolidates the top 2 in. of soil, the reader will have little difficulty in seeing that the implement can rarely, if ever, have an appreciable effect in bringing water upwards. What it does is to make the soil firmer in the surface inches, thus giving a better contact between the soil particles and the seed and young roots, which will therefore be better able to obtain what moisture is present.

As a corollary of the capillary rise theory, there was a satisfying explanation of why water is conserved in the soil by keeping the surface in a loose condition; the view was that the capillary tubes were broken in the surface mulch, and so water did not rise quite to the top, and thus escaped evaporation. There seems no room for doubt that a surface mulch does conserve moisture, but the above explanation of the phenomenon can no longer be accepted. Presumably when the mulch has dried out it checks the diffusion of water vapour, produced lower down, into the atmosphere. Certainly a loose surface ensures that rainfall will be absorbed into the ground rather than be held on the top and largely lost through evaporation. This last point is important because the loss of rain water, by evaporation before it has sunk in, is considerable; it may be claimed as a general benefit of tillage that the loss is reduced because water soaks readily into tilled land. Through unmoved clay, water can only percolate down fissures and cracks, but cultivated soil acts more like a sponge, and, as it were, mops it up. Tillage increases the air spaces in the soil and so raises the water-holding capacity; this is usually beneficial but, on the other hand, as will be seen later, it may be disastrous.

In tillage an important object is to form the soil particles into little aggregates, or crumbs. On light land this condition is required because otherwise the soil will not hold water, and also, if it is a mere powder, it may blow away in a high wind; but on sandy land crumb formation is more dependent on the humus content of the soil that on cultivation. The majority of soils contain some clay, and the higher their clay content the more important, and the more difficult, it is to produce the crumb structure; in that condition the soil may be firm and yet easily permeable by roots, whilst water will pass through it, air being drawn in behind the water. The great agent in reducing clay land to the mellow crumby condition is the weather. Frost is the most potent weather factor, the expansion of water on freezing having a great shattering effect on clods; whenever possible the heavy-land farmer leaves his land in plough furrows over the winter to obtain this effect at its maximum. In summer the heating and cooling of day and night disintegrates the clods, owing to unequal expansion and contraction of their constituents. A less understood, but very powerful, factor is that of wetting and drying, which has a truly remarkable effect on clay. Heavy land ploughed up in a wet condition is quite impossible to work down to a tilth until it has dried through; at its first drying it consists of flinty hard lumps, but then after rewetting its amelioration appears to the uninitiated almost miraculous. In its original wet condition the spaces in the clod were filled with water, which served to bind it together, but after drying and rewetting the spaces have air imprisoned in them, and these pockets form points of weakness at which shattering will occur; furthermore, there is, with each drying and wetting, a change in the physical state of the clay colloid, which favours disintegration of the clods. If the drying and wetting are repeated several times the clods will break down when subsequently struck by the tine of an implement. Similarly, clay land which is lying hard and dry will not work down until it is rained upon; as it is usually expressed, the rain 'softens' the clods. Of course clay soil which has received the benefit of this weather factor should not be worked when wet, or deflocculation will occur, and it will be returned part of the way back to its original sticky condition; on the other hand, it is not much use working it when it is very dry, because then the lumps will not shatter. The ideal condition for working can only be recognised by experience, but roughly it is when the land is dry enough to prevent it picking up on the boots, and yet containing sufficient moisture so that the clods break down readily when kicked.

Tillage must be viewed as a constructive process—the building up of soil crumb. This is produced by long exposure of the upturned furrow to the weather, followed by working of the soil in the correct half-wet, half-dry condition, and the crumb will persist for more than a year as long as the land is not worked when wet. Much land that was reclaimed in the War of 1939–45 proved difficult to work at first but where the exigencies of the situation permitted a fallow before the land was cropped this difficulty did not arise; by 'getting the sun into the land' crumb was formed and then good seedbeds were obtained. Recent years have seen a great and very regrettable spread of disk cultivation in this country; the disk has a destructive, not a constructive, action, the principle being to cut the soil up into little bits. This produces an imitation seedbed, level enough on cursory examination but sadly lacking in crumb. On clay land disks often prevent crumb formation by leaving behind them cut surfaces which are 'puddled'; the unkind, knobbly seedbed, with no large clod but no mould either, is quite characteristic of the disk. A disk runs well behind a tractor and keeps itself clean of rubbish but that is not sufficient justification for it. There is, however, one case in which the disk is supreme and that is where a ploughed-up turf has to be worked to a seedbed; the fibre in the turf prevents puddling and the ruthless action of the disk is necessary where thick sods have to be reduced. A further advantage of the disk is its quite surprising effect of making the land firm; a newly turned furrow may lie 'hollow' and nothing corrects this condition better than the disk.

Aeration of the land is one object of tillage and it is difficult to decide how much importance should be attached to it. There is no doubt that plant roots and beneficial bacteria require oxygen and cannot grow in soil which includes no air. Plants die in waterlogged land because the spaces between the soil particles are filled with water to the exclusion of air; good drainage permits the excess water to seep downwards and as it does so air follows to fill the vacated spaces. Apart from water-logging it must be rare for conditions to arise which lead to the total exclusion of air from the soil but there may well be small patches of the soil, even in the surface inches, where the passage of air is seriously impeded. Clay soil may 'run together' in impermeable clods and even on silty land, if organic matter is deficient, the same sort of conditions may arise; plant roots will not permeate into such localities, possibly because they are mechanically impeded but more certainly because of the absence of oxygen. There is also the point already mentioned in connection with fallowing, that aeration is necessary for the conversion of plant nutrients from the unavailable to the available form; frequent cultivation would therefore seem desirable in order to expose fresh soil to the full effects of aeration. Experiments have been conducted to compare generous and niggardly cultivation in seedbed preparation, and the usual result has been that the yield of the crop was unaffected. But a word of warning is necessary here, because all the experiments have been short-term ones, and as yet no results have been published of experiments comparing long-continued over- and under-cultivation; the claim that a policy of thorough cultivation, within practical limits, justifies itself by rendering plant food available cannot therefore be regarded as disproved. It does at least seem probable that ploughing, by bringing a fresh layer of soil to the surface for weathering, improves not only the physical, but also the chemical, condition of the soil; but to the practical man the point is hardly worth establishing, since the plough, as a means of loosening the soil, and of burying weeds and rubbish, is not likely to lose its pre-eminent place as the first implement of tillage.

The control of insect and fungoid pests might be claimed as a final object of tillage. By turning the land over and stirring it insect pests are exposed to the birds, which can often be seen busily engaged behind implements; no doubt considerable good is achieved in this way, especially on land infested with wire-worm. It is not possible to cite many cases of pathogenic fungi which can be controlled by cultivation, but take-all in cereals might be quoted. Early ploughing of infected stubbles after harvest is recommended as a measure to be taken against this disease, and there is some evidence that it may be checked slightly by harrowing wheat in spring.

The above brief and discursive account admittedly gives but a nebulous impression of the objects of tillage. Ideas on the subject are now in a state of flux, and experimenters have not succeeded in disentangling the essential from the non-essential requirements. That weeds must be kept down, rubbish and manure buried and sufficient mould produced to cover the seed, cannot be disputed; that the soil should be inverted by the plough once in preparation for sowing seems highly probable. Further than this all is conjecture. The traditional farmer claims that the thorough working he gives is in the nature of an insurance against an unfavourable season, but what risks he is covering, and what premium it is reasonable to pay, are matters on which he is ignorant; and, unfortunately, no one has the necessary knowledge to instruct him. Further general discussion would therefore be unprofitable, and it is proposed to turn now to the actual requirements of a seedbed, a subject closely allied to the above, but one on which ideas may be crystallised rather more definitely. The requirements will be discussed under five heads, and no effort will be made to arrange these in order of importance.

(*a*) *Clean.* Little more need be said on this subject, but it is very important that a seedbed should be reasonably clean. In the case of root crops it is folly to sow on land infested with perennial weeds, because the hoeing which the crop subsequently receives is ineffective against these species; very little experience is necessary to convince one of the absurdity of hoeing established couch. Although annuals may be killed by the hoe it is good practice to kill as many as possible before sowing the crop, so that the young plants may not suffer very severe competition as they emerge from the ground. For cereal crops a clean seedbed has become more important as economic exigencies have reduced the amount of hoeing performed. It is bad farming to sow in an unclean seedbed, and this applies to all crops with the possible exception of potatoes, for which after-cultivation is intensive and can be started early and continued for a long time; even in that case couch may prove very troublesome.

(*b*) *Rich.* The seedbed should not only contain an adequate supply of plant food, but should be mellow—that is, the plant food should be distributed reasonably evenly through the soil. If farmyard manure is ploughed in during the autumn for roots and the field cultivated and drag-harrowed in spring, the lumps of manure will be disintegrated, and the plant nutrients well mixed with the soil. Chemical fertilisers should be applied before the final cultivations, which will work them into the ground. Whether this mellowness is important in all cases is somewhat open to doubt. In its absence some plants will find a rich pabulum in their immediate vicinity, whilst others will be half-starved, but the favoured ones may flourish sufficiently to compensate for their weaker brethren; it is, therefore, impossible to say what degree of mixing of plant food in the soil is desirable, but thorough distribution is certainly advantageous for barley, in which even growth is essential for a high-quality malting sample.

(*c*) *Moist.* It is very important that a seedbed should contain enough moisture to ensure rapid germination. Seeds which lie long in the soil may lose their vitality and ultimately establish weakly plants, whilst some will 'chit' and then, finding insufficient moisture, die; furthermore, plants are subject to pests whilst emerging from the ground, and it is desirable that they should remain in that vulnerable state for as short a time as possible. On the other hand, the seedbed should not be too wet, because then the soil will be puddled and will pick up on feet and wheels, which will mean an uneven sowing depth.

It is very common for the land to be in a dry condition in early autumn, and then the heavy-land farmer is in rather a quandary. If he decides to drill in early October, despite the dry soil conditions, he is dependent on rain falling in the very near future, and in many cases an unsatisfactory crop results. If, on the other hand, he decides to wait for the rain, he is running the risk that the weather may break completely and a prolonged wet period ensue. Most farmers prefer the second risk, the feeling against sowing in dry soil being very strong; it will also be realised that when heavy land is dry at that time it is often in a 'nobbly' condition, with very little crumb, and that rain is needed to soften it before a good tilth can be obtained. With spring corn the reverse is usually the case, the common experience, particularly on heavy land, being that the farmer has to wait for the land to become dry enough to permit drilling; in some discouraging seasons he may wait in vain and have to 'maul' the seed into wet ground, or lose the crop altogether. It is quite common for winter corn to be 'mauled' into the ground and yet to produce a good yield, but the same liberties cannot be taken with like impunity in the case of spring corn; in the latter case there is no hard weather to come and less time for natural agencies to correct the imperfections of the seedbed. With root crops, and more still with seeds leys, the chief danger is that of 'malting'; that is, the seed is put into dry ground, starts to grow and then dies for lack of moisture. The danger is a very real one, as the seed is small and must be sown very close to the surface; all that can be done when conditions are dry is to mitigate the risk by sowing just a little deeper than usual, and to follow with a roll, so that the soil may be pressed close to the

FIGURE 2

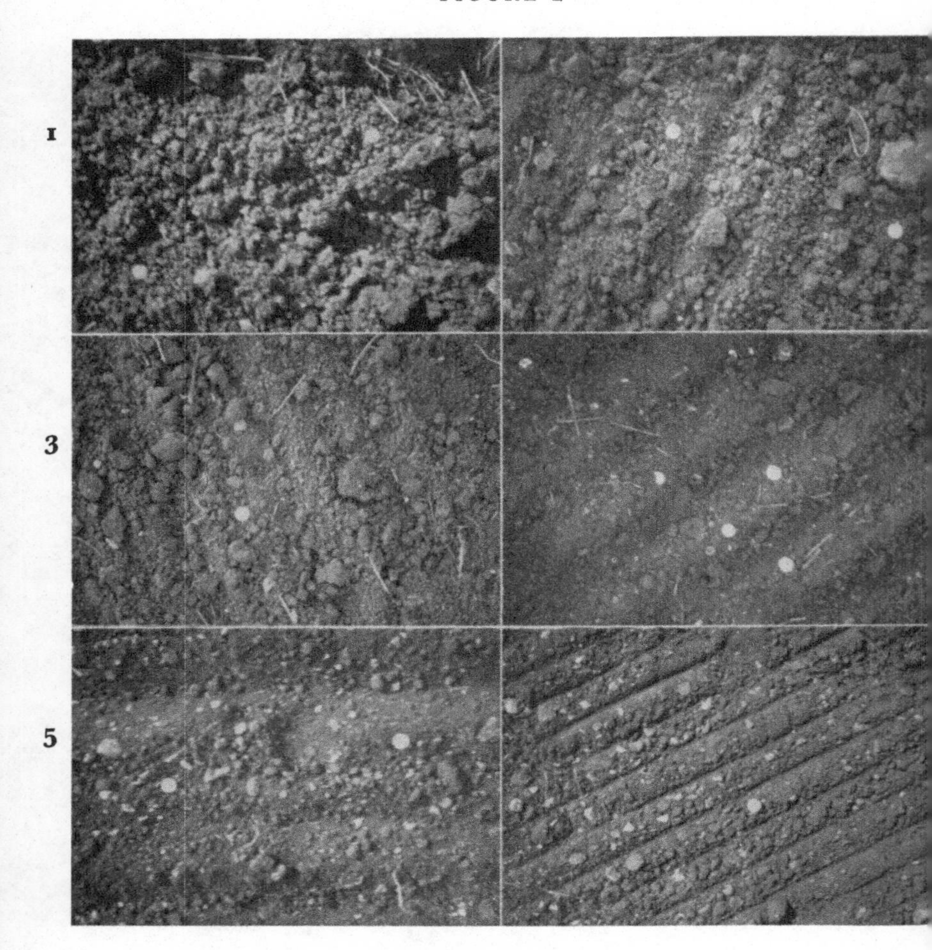

DEGREES OF TILTH

1. Rough autumn seedbed; level but deficient in crumb.

2. Ideal autumn seedbed.

3 & 4. Seedbeds for spring corn. Note much crumb, no large clods.

5 & 6. Seedbeds for sugar beet. Note better conditions for even depth of sowing after ring roll.

(Coins are half-crowns; diameter $1\frac{1}{4}$ in.)

seed. With all crops the actual date of drilling is very highly dependent on rainfall; there is great scope for judgement in the matter and no little hangs on luck.

(*d*) *Tilth*. If the surface is very rough the seeds cannot be drilled and covered to an even depth. On a rough seedbed some will not be covered at all, and may be eaten by birds, or give rise to plants which may die in a dry time when they are still young; others fall into large holes and either fail to get through at all, or produce yellow and weakly plants. The fineness of tilth required varies according to two factors—the size of the seed sown and the time of the year of sowing. A small seed contains a small reserve of food, and therefore cannot establish a plant unless it is sown near the surface of the ground; shallow sowing necessitates a fine, even seedbed, so that there is plenty of crumb to give a thin layer of soil above the seed, and an absence of clods, one of which will give too thick a covering. Therefore the rule is—the smaller the seed, the shallower it must be sown, and, hence, the finer the seedbed it requires. As regards time of year the difference lies between autumn and spring or summer sowing. A fine seedbed in autumn is undesirable for two reasons. In the first place a fine autumn seedbed will be further reduced by the winter frosts, and then beating rain may make it, if it contains much clay, run together and set when it dries, so that the surface carries a hard pan; this will prevent rain from penetrating into the soil and, in a bad case, may even injure the plants mechanically, either by shrinkage on drying, by pinching the stem, or by lateral movement if the field is harrowed. In the second place, surface clod provides shelter for the young plants, by breaking up the cold winds that blow over the ground. It is a common observation that autumn corn when sown on a rough seedbed looks greener and healthier in spring than when sown on a very fine tilth; crumb is required to ensure that the seed lies 'snug' in the soil, but a good sprinkling of clods, of the size of a man's fist or a little larger, is also required. The extreme difference would be between the tilth required for beans in autumn and for clover in spring; the former should be rough and cloddy, the latter smooth and all mould.

(*e*) *Depth*. Plant roots often penetrate to a depth of 6 ft. and more, despite the fact that it is usually only the top 6 in. or so that is tilled; but by washing the soil away from the roots of growing plants it is found that only a few roots permeate the unmoved subsoil, whereas in the worked layer there is a thick network. Such studies have shown conclusively that root systems are highly affected by the condition of the soil and that proliferation is more abundant in layers that have been stirred than in solid ground. From this it would appear that deeper cultivation, by extending root range, must benefit the plant, and now that mechanical power is available deep working presents no insuperable difficulties. But the results, in practice, of deep cultivation, have been very variable, and this is scarcely a cause for wonder in view of the different classes of soil and subsoil on which it has been tried, and of the different methods adopted.

Deep ploughing is the usual method of extending the tillage depth. There are many situations, where there is a deep soil so that the material brought to the surface is not very unlike the normal top soil; silts, fens and loams often provide these conditions, and on them ploughing to a depth of 18 in. or more has become a common practice—perhaps adopted once in a rotation—of some farmers. But in the majority of cases there is a very clear demarcation between the top, worked layer, of about 6 in. depth, and the subsoil below it; the former is not only in better physical condition but contains much more humus, as can be seen from its darker colour. To bring to the surface sticky clay which has been undisturbed for centuries is to court trouble. This material is practically devoid of plant food and is very 'unkind'—that is to say, it will not work down into a tilth; even a winter's weather will not produce crumb, the tendency being to form flakes. And it is not only on clay land that deep ploughing may lead to difficulties; on light land the material brought to the surface may be nearly pure sand or gravel, inert material that merely dilutes the useful top soil. Deep ploughing must therefore be adopted with circumspection; it may be very beneficial, but where the subsoil is markedly different from the topsoil depth should be increased slowly, say an inch or two at a time, and the farmer should satisfy himself that the results are promising before proceeding to extremes. There are many farmers who would not agree with this cautious advice, who plough deeply on all sorts of soils, not even excluding the thin chalks. These are generally men who farm well; they get good crops but it is impossible to disentangle deep ploughing from other factors such as generous manuring as a contributor to their success. Only definite experiments covering many different soils and situations can test the matter and in 1944 Dr E. W. Russell* started such a series of experiments on private farms scattered widely over England. In these experiments plough depths of 12–16 in. were compared with depths of 6–8 in. and results were very variable. Thus with forty-seven potato crops deep ploughing raised yield by at least ½ ton an acre in 40 per cent of the crops but lowered it by a like amount in 20 per cent. Similar inconsistencies were found in twenty experiments with sugar beet and in 44 with cereals and the results gave no clear indication as to which soils would, and which would not, respond favourably to deep ploughing. In general, results for deep ploughing were more favourable on heavy than on light soils and the conclusion was drawn that for the best results deep ploughing should be accompanied by very generous manuring. Although cases occurred where many weed seeds were brought to the surface by going deeper the more common effect was that deep ploughing gave useful control of weeds. Where land is ploughed deeply in autumn it is usually possible to work it earlier in spring than after normal ploughing; in experiments on private farms advantage cannot be taken of this since the whole area has to be treated alike, after the experimental ploughings have been done, to keep dislocation of normal practice to the minimum. Further experiments are therefore being done on Experimental Husbandry Farms where it is possible to cultivate and sow plots separately so that any benefit of earlier spring work after deep autumn ploughing may be reaped; shallower ploughing than 6–8 in. is also being tested. Up to the present these experiments have given no clearer indications as to the value of deep ploughing than did the earlier ones.

Subsoiling differs entirely from deep ploughing in that the object is to break up the subsoil without bringing it to the top. It was performed with horses, when the common procedure was to use a subsoil plough which followed an ordinary plough and

* 'Effects of Very Deep Ploughing and of Subsoiling on Crop Yields.' E. W. Russell, *J. Agric. Sci.* **48** (1956), 129.

burst up the bottom of the furrow; the loosened subsoil was covered by the next furrow slice turned. But with high-powered tractors the two operations can be performed at once; behind each plough furrow there runs a strong tine which bursts the subsoil to a depth of 6 in. or more below the plough furrow. This operation is directed particularly against hard pans which not infrequently occur, either as a result of continual ploughing to the same depth (that is the pan is formed mechanically by the heel of the plough) or by deposition of salts from solution in the water draining through the soil. Subsoiling is quite common on light land in which pans frequently arise from the latter cause, but on heavy land it is often ineffective, as the subsoil is rarely dry enough for any bursting effect to be obtained; if a subsoiling tine be drawn through wet clay the cut it makes is apt to close up again very shortly afterwards. On the whole, experimental evidence available is favourable to subsoiling, though cases have been reported in which crop yields were apparently reduced by it; it may be that in these cases the subsoil was left in too open a condition, and that the harm was caused by too free drainage or by poor root hold. In some cases results have been wholly favourable to subsoiling with a variety of crops, but more particularly with sugar beet; the breaking of a pan below the normal plough depth not only increases the yield of beet, but leads to a marked reduction in the proportion of fangy roots. It is quite impossible to speak generally and say that subsoiling is either good or bad; it is an operation which, at least on the lighter soils, is worth trying; where a farm appears to benefit it should be performed occasionally, for instance, once in a rotation. It should be mentioned that the steam cultivator was sometimes used as a means of deepening the tilled layer. The power available was sufficient to break up even sticky clay land to a depth of 9 or 10 in., and some clay-land farmers made it a principle to steam-cultivate their fallows, so that the land might be stirred slightly below the normal depth of working. Nowadays heavy tracklaying tractors take the place of steam power and can be used for the deep cultivation of fallows.

In 1932 an implement of great apparent promise for deep cultivation was introduced to British agriculture. This was the 'gyrotiller', a very large and powerful machine whose working parts consisted of strong skyves for stirring the land horizontally to a depth of 18 in. or more. For several years 'gyrotillers' were used fairly widely but farming opinion, which was at first favourably disposed towards them, swung in the opposite direction; this was in agreement with the results of controlled experiments which failed to show any advantage in crop yield following the operation. There were some unhappy cases where the deep stirring of the 'gyrotiller' resulted in a condition of the soil resembling a quagmire. It produced a tilthy condition in the subsoil which absorbed much water in wet periods; sometimes the implement disturbed drains and then the deep tilth became a slithery mess which was very slow to dry so that a field might remain impossible to cultivate for as much as two years. In the War of 1939–45 the 'gyrotiller' gave yeoman service in reclamation work but it has no place in normal farm practice.

There is no simple answer to the question as to how deep the soil should be tilled; the answer depends on the soil and subsoil, particularly the latter. The farmer's normal aim is to work to the 'full ploughing depth'—that is to the point where the unweathered, unmoved subsoil is reached. Before going deeper he should spend some time on the field with a spade, exploring what is beneath. If there is no clear demarcation between the soil and the subsoil, deeper working will probably be good policy as giving greater root range; but usually there is a change in appearance from soil to subsoil, and then wariness, rather than boldness, is to be advised.

(*f*) *Firmness.* It has been said that the land should be loosened to a depth depending on the amount of weathered soil, so that roots may penetrate more easily; having been loosened, it should be rendered firm again before seed is sown. It must be fully realised that the firmness obtained after the land is worked is very different from the solid condition prior to working; in the former the structure is crumby but the crumbs are pressed close to one another, whereas in the latter the particles have run together and the land is set hard. One reason why firmness is desirable is that otherwise drilling to an even and proper depth is impossible; wheels and horses' hoofs will sink into the ground, and the drill coulters will cut in to a considerable depth even when no press is applied. But the main reason against drilling on too loose a tilth is that a high proportion of the space occupied by the soil will consist of little pockets of air, and that a seed will often lie bridging the gap between two crumbs; if the soil is firm the gaps will be smaller, and the seed will be in more intimate contact with the crumbs, from which it will be able to absorb moisture. And what is true of the seed is also true of the roots later in growth; the root hairs, which serve as the feeders of the plant, are very minute, and in a loose tilth they may be lying in air over a large proportion of their length. The aim is to provide conditions so that there are gaps between the soil crumbs, enabling the rootlets to force their way, but that the gaps are small so that contact may be nearly continuous. This will clearly make for efficiency in absorbing moisture and plant food present, and it will also serve to anchor the plant firmly to the soil; the latter is important with cereal crops which are liable to lodge, lodging in some cases consisting of a heeling over of the whole plant. Firmness is more important on light than on heavy land. The latter is very apt to become too firm, either through being worked in wet weather or through the action of rain; both of these may deflocculate the crumbs and tend to bring the soil back to its original condition. On light land many of the soil particles are large, and the aim is to bind them together with the colloidal material (humus and what clay particles are present) and to press them as close together as possible. Consequently the roll is used much more on light than on heavy land, its main function on the latter being that of crushing clods.

But the roll is an unsatisfactory implement of consolidation. On a loose tilth the surface inch or two take all its effect and act as a cushion, so that little firmness is produced in the lower layers. The general method of seedbed preparation is to start with the deeper working implements and finish with the shallower working ones; of course, in getting a seedbed it is not usually necessary for the farmer to use all the implements at his command, but the general sequence is plough, cultivator, heavy harrows and light harrows. The important point to appreciate is that the later, shallow-working implements not only stir the surface inches, but at the same time they consolidate the ground below; the weight of the implement helps in this, and the vibration causes soil crumbs to nestle down in the spaces beneath them. This consolidating effect of shallow-working implements is important and is expressed in the farmers' saying 'The harrow is the best roll.' In addition to the effect of the implement there is the compressing action of the horses' hoofs or tractor wheels; if the field is traversed four or five times a large proportion of the ground will receive this squashing, which is altogether more intensive than the gentle pressing effect of a roll.

In general there are time intervals between the various cultivations in preparing a seedbed so that some weathering (at least some drying) occurs; this is desirable, as it favours crumb formation. With high-powered tractors it is possible to assemble a 'train' of implements to draw across the field, with the object of preparing the seedbed in one fell swoop; this saves labour and gratifies a mechanically minded farmer's sense of power, but the whole gamut together are considerably less effective than when used separately with intervals, possibly only of a few hours, between them. It will also be appreciated that where a turf or rubbish is buried it is desirable to allow a considerable interval between ploughing and subsequent cultivations, so that the buried material may die before any of it is brought back to the surface. Engineers have striven, with some success, to evolve single implements which will form a seedbed from solid ground at one operation. The rototiller breaks up the ground effectively and gives sufficient tilth, but the seedbed it produces is not favoured by practical men; the field is left in a 'fluffy' condition so that when walking across it a man sinks in over his ankles, and rolling, however heavy and repeated, does not give that loose surface and firm bottom which are required. In fact the roll tends to give a firm top and loose bottom to the seedbed, and the experience is that seeds sown in such conditions grow well, but that the plants fade away at the first sign of drought.

RIDGE *VERSUS* FLAT

Potatoes are always grown in the ridge, because only so can they be given the loose soil in which the tubers can grow and remain covered by soil, and yet be easily removed from the ground. Root crops can also be grown on the ridge, and that is the common practice in some districts of the country, whilst in other parts they are normally grown on the flat; it is interesting, and important, to enquire why there should be this difference in practice. The advantages of growing root crops on the ridge are as follows:

(*a*) *Early hoeing.* From the outset the position of the rows is known and hoeing can therefore be started before the crop is up. This is an important advantage, especially in wet districts where weeds appear rapidly and profusely, and on land that is foul.

(*b*) *The crop is above the weeds.* As the plants emerge from the summits of the ridges they start level with the most favoured of their competitors, and considerably above the majority of them.

(*c*) *The texture of the soil in the ridges is not lost in wet weather.* A fine seedbed, such as is required for root crops, is very apt to run together under beating rain and in waterlogged conditions. The tops of the ridges, at least, will be well drained and dry, unless the whole field becomes completely flooded. The point may be supremely important in a wet season, as the dryness of the ridge tops may permit seeding much earlier than would be possible on flat land.

(*d*) *The best soil, and, if desired, the manure is concentrated under the plant.* This is probably important in the early stages of growth, though, with sowing on the flat, the plant roots will soon join across the rows and thoroughly explore all the soil. In the later stages the 'bulb' will be swelling, and where sowing was on the ridge it will be in loose soil, and encounter less resistance than when sowing was on the flat.

The disadvantages of ridge work are as follows:

(*a*) *Loss of moisture.* This is the great drawback, moisture being lost in two ways. In forming the ridge the surface soil tends to go to the centre of the ridge, whilst soil from below the original surface goes to the outside; thus just before drilling the damper soil is exposed to evaporation, and in a dry season this is the last thing the farmer wants. Secondly, the surface area of the field is increased, so that there will be more evaporation throughout the period of growth.

(*b*) *Extra working.* This is not very serious, but the actual ridging of the field is an extra operation to be performed after a suitable tilth has already been obtained.

(*c*) *Clods tend to be collected in the middle of the ridges.* There should, of course, be no large clods on the field at the time of drilling, but such as are present tend to roll down the side as a ridge is formed, and the completion of the ridge, the next time across the field, covers them. Too much clod in the middle of the ridge makes it open and more susceptible to drying.

(*d*) *A special drill is needed.* It is possible to sow on the ridge with an ordinary drill, but it is difficult and crop rows are apt to appear half-way up the ridge. The turnip drill is not an expensive implement, however, and this point must not be allowed too much weight.

(*e*) *Hoeing tends to draw the soil away from the plant, leaving the 'bulb' exposed.* This point is only important in the case of sugar beet, and can be met by the use of the double-mould board plough after hoeing is completed.

The balance of the argument is in favour of the ridge in a wet district, since the loss of moisture is generally unimportant and weeds grow very rapidly. In a dry district, however, the balance swings the other way, since rainfall is normally the factor which limits the yield of root crops, and any avoidable loss of moisture is greatly to be deprecated. These conclusions are in accord with practice, because it is in the wetter districts of the country that roots are commonly grown on the ridge, whilst in the drier districts they are usually grown on the flat. If a line be drawn from the Humber to the Bristol Channel, it will roughly coincide with the division in common practice; to the north and west of the line ridge work may be said to be the general rule, whilst to the south and east roots are usually grown on the flat. Not that the line must be regarded as rigid, because individual farmers have their own predilections and many cross the line, taking with them the methods to which they are accustomed. It is worth noting that the common methods of practical men conform sensibly to the conditions, and that each method is right in its own home.

CHAPTER V

THE PREPARATION OF THE SEEDBED

The general principle of preparing the seedbed is that a layer of earth is first separated from the top of the unmoved land (usually by the plough), and, after as long an interval as possible for weathering, the separated layer is reduced to the required tilth, deep-working implements being used first and shallow-working implements later. But no useful rule of thumb can be enunciated, nor can a precise sequence of operations be given, even for rigidly defined conditions of soil and cropping, because the farmer is always at the mercy of the weather. To the infinite variation caused by soil and weather differences must be added that arising from the partiality in individual farmers for certain implements. In the south-west of the country the spring-tined harrow is the jack-of-all-trades in tillage, and it performs yeoman service in all sorts of conditions; but in the eastern counties farms usually include a larger acreage of arable land and carry a considerable collection of implements, a collection in which, surprisingly enough, the spring-tined harrow rarely finds a place. Reference was made on p. 135 to the disk, its advantages and its failings; there are available disk harrows of various weights, disk drills and even disk ploughs. The objection to them is that they force a sort of tilth, an unkind and knobbly affair, rather than build up a good crumb structure in conjunction with weathering. They are, however, in tune with modern methods and enable a farmer to push on with his work, albeit that the final result may be imperfect. They have undoubtedly come to stay in British agriculture but they are unworthy of the high place accorded to them by some farmers; the man who ploughs a field and after a decent interval just 'disks it down', irrespective of the infinite variety of conditions that may obtain, is reducing cultivation to a routine which ignores the opportunities offered in each particular case. Similarly the rotovator may be a powerful aid in cultivation but it should not be used without due thought; on solid ground the backs of its flying blades may produce a hard, panned condition which checks aeration and may prevent root penetration. Certainly some differences between the implement equipment of farms are adequately explained by variations in size of holding or in soil, but it is also true that many are not; as long as tillage remains an art its paraphernalia will depend to some extent on the artist's temperament, and personal idiosyncrasies cannot always be justified in the cold light of reason.

It is quite impossible to describe all the methods—even all the common methods—by which seedbeds are prepared. In what follows the given sequences of cultivations are to be regarded as merely typical; it must also be remembered that the actual date of each operation can only be determined by the man on the spot, who must use all his art in deciding when the land is in the right state, and what implement will best effect his purpose. The dates given below are subject to this very important reservation; they are given to indicate roughly the time relations between the operations. The common cases may be classified for discussion under six main heads.

I. AUTUMN-SOWN CROPS

The chief crop to be considered in this section is wheat, but the other cereals and beans may also be included. There is no appreciable difference in the seedbeds required for these cases, though for beans, which are usually sown rather deeper than the others, the ideal seedbed would be somewhat rougher. In all these cases some clods are desirable to provide shelter for the young plants during winter, and on heavy land this is further important in order to prevent the surface running together; but, though clods are required, mould is also necessary, so that the seed may be in close contact with the soil. Where the land is broken up in good time the seedbed should be very easy to obtain, but on clay land, if the original ploughing is late and carried out when the soil is wet, the condition at drilling is often very rough; it is surprising how these crops, and wheat in particular, sown in an extremely rough seedbed, sometimes eventually turn out very satisfactorily. A farmer who is well up with his work may, with the greatest advantage, seize the chance offered by a dry September to cultivate or drag-harrow his fields in that month; valuable cleaning may thereby be effected in a favourable season, but as the opportunity occurs rarely such tillage is not included below.

(1) AFTER FALLOW

The crop sown is usually wheat, but occasionally oats, and the case is a very simple one, particularly if the last operation of the fallow was to plough immediately after harvest. The field will then be left until October, when a light harrow to smooth out the furrows (which will be very broken after the summer working) will often be all that is necessary to do before drilling the field; in some cases a large number of weed seedlings appear after the final ploughing and then sufficient and heavier harrowing should be carried out to kill them. If September is wet the field may settle down to a 'sad' condition and a drag harrow, or even a cultivator, may be advisable to loosen it; on the other hand, if the September is very dry additional harrowing, or even rolling, may be desirable, because the condition will be too loose. After a bastard fallow the tilth will not be so fine as after a full fallow, and one or two strokes with a drag harrow may be given to reduce the clods. These fields will be the earliest sown ones, and particular care should be exercised that sufficient moisture is present at the time of drilling; often it is necessary to wait for rain, but the first favourable opportunity for drilling should be taken, because of the danger that the fine tilth may absorb much water, and prove very slow to dry, if the weather become very wet.

Where mustard has been grown at the end of a fallow (usually on light land) it is ploughed in, probably after rolling it down in the direction of the ploughing, in October. A good man and a well-set plough with drag chains are required to cover it all in satisfactorily, and the plough is usually followed by the roll, or by a furrow press. No deep harrowing will be done before drilling, because of the danger of dragging the mustard back to the surface, where it will be pushed along in front of the drill coulters.

(2) AFTER BEANS

The land may be assumed to be heavy and the succeeding crop is nearly always wheat. Beans are widely credited with 'leaving the land well for wheat', and are, in fact, sometimes grown principally because they are a good preparation for wheat. There are several reasons for this. A good bean crop will enrich the soil with nitrogen and, if well hoed in the spring, will also leave the land relatively clean; furthermore, the strong tap roots of the beans will break up the furrow slices, and the spring hoeing will have provided a surface mulch to check the drying out and cracking of the land in the summer, so that a tilth will readily be produced.

The field should be ploughed as soon as possible after harvest (some delay so that sheep may be run over to pick up the shed beans is advisable) and left to weather until the date of drilling. 'Wheat likes a stale furrow' is a saying to which nearly all farmers will subscribe, though it is one for which experimental proof is still lacking. If the land is left for some time in the ploughed state the furrows will settle to give a firm bottom to the seed bed ('Wheat likes a firm bottom' is another farming belief), and crumb will be formed by weathering; it is certainly true that the seebed will be produced more easily if the furrow is stale. One stroke with the drag harrow is often sufficient to produce the necessary tilth, though if the soil is a little 'unkind' this may have to be repeated or followed by a light harrow. These harrowings are carried out just before drilling, and it is the common practice to have all the implements—say, drag harrow, light harrow, drill and seeds harrow—on the field at the same time. It is important to avoid leaving heavy land in autumn for any length of time in a state ready for drilling, owing to the danger of heavy rain. It is the normal practice to put the light seeds harrow over the field twice after the drill to ensure a good cover of the seeds; the second time the harrowing is across the drill rows and, of course, this cannot be commenced until the drilling of the field is completed. Thus the complete operations might be:

Plough	September
Drag harrow Light harrow Drill Seeds harrow	October 20th
Seeds harrow (across)	October 21st

(3) AFTER A CEREAL

Generally the crop to be sown is winter oats or beans, but it might be winter barley or wheat. On heavy land the procedure would be very much as in the preceding case, except that the land will not work so readily; it may therefore be necessary, having ploughed it and left it as long as possible to weather, to use the cultivator once or twice to break up the furrow slices. If beans is the crop to be sown, farmyard manure will usually be applied before the ploughing, and then the cultivator is better withheld, lest its deep tines bring much of the dung to the surface again.

On light land, where the case also occurs frequently, there is little difficulty in obtaining the required degree of tilth, but great importance is attached to getting the land firm. The roll, therefore, often follows the plough. The effect of the roll immediately after ploughing is very different from its effect on loose, tilthy land; in the former case the furrow slices do not disintegrate under it, and are pushed down as units, so that consolidation is achieved at the bottom of the furrow, whereas in the latter case it is only the surface inches which are rendered firm. Many light-land farmers insist that the roll should follow the plough very closely, if possible on the same day; due weight must be allowed to the practical man's belief in the efficacy of this, although experiments have not yet succeeded in detecting the benefit. On light land there is no great risk in leaving the furrows rolled down for some time, and drilling may be direct on the rolled surface, or be preceded by a light harrow to loosen the top. Thus on light land the procedure might be:

Plough	September 20th–30th
Ring roll	September 20th–30th
Light harrow Drill Seeds harrow	October 30th
Seeds harrow (across)	October 31st

(4) AFTER A SHORT LEY

Wheat nearly always follows a short seeds ley, though some farmers prefer to graze their field till late in the autumn, and follow with a spring-sown crop. After a seeds ley the land should be in good heart and clean; if the ley contained any appreciable amount of perennial weed it should have been broken up about midsummer, and a bastard fallow taken. The land will be very solid, not having been moved for at least 18 months, and will carry something approaching a turf. If the field can be ploughed early there is no particular difficulty about the seedbed, but if the ley is wanted for grazing until the end of September, or if a crop of red clover seed is not finally cleared until early October, there is little time for weathering and settling. Ploughing should be as early as possible, care being taken to cover up all the green material; skim coulters will generally be required, and if there is much rubbish or green growth on the surface, drag chains as well. The ploughing should not be very deep because of the difficulty of correcting hollowness if the depth exceeds 4–5 in. The major difficulty presented by this case is the fact that the solid furrow slices, bound together by the roots of the clover or other plants of the ley, will rest unbroken against each other, with quite large spaces beneath them. This hollowness of the bottom of the seedbed is regarded with the greatest displeasure by farmers, as it is believed to lead to a poor root hold in the following crop, with consequent under-development and liability to lodging. Different farmers treat this danger with varying degrees of seriousness, and consequently there are several methods of procedure commonly found in practice.

In some instances no special steps are taken to correct the hollowness. In that case, when the furrow slices have been lying exposed to the weather as long as possible, the field is harrowed, probably with a light harrow, several times. To break the crests of the furrows most thoroughly the harrowing should be across them, but this may tend to roll them over, bringing the buried, but still living, green stuff to the top again; the best procedure, therefore, is to harrow along the furrows once or twice and then across them. If the field be harrowed several times considerable consolidation will occur, and over much of the field the slice will be crushed down into the hollow beneath it by horses' hoofs or tractor wheels. In the north of the country it is a common practice to broadcast the seed before harrowing, which will serve to cover it in, but in the south and east the harrowing is usually done first, and the seed then drilled and harrowed in.

One method occasionally used to overcome the hollowness at the bottom of the seedbed is to break the original surface before turning the land over. If the field is first disk harrowed twice the loose soil will fall into the hollows on ploughing, but the surface is often so firm that the disk harrow has little penetration.

There are three methods of pressing the furrow slices at the time, or soon after, they are turned. As was observed in discussing the previous case, a heavy ring roll immediately after ploughing will do much to make the bottom of the seedbed firm, and this implement is very often used at this stage. A more effective implement is the furrow press, which is often attached to the plough now that tractors provide sufficient power for the double load. Each large rim of the furrow press runs at the point where one furrow leans on its neighbour, and the weight is sufficient to push the slice down into the hollow below. The third method, though it might appear archaic, is not very expensive and is very effective. It is to plough with a single-furrow plough, and to follow the plough with an ordinary farm cart, driven so that one of its wheels runs on the freshly turned furrow. This might be judged very wasteful in labour—to employ a driver, a horse and cart for the sake of the compression given by one of the cart's wheels—but the wheel tracks obliterate the hollow below the furrow and provide an admirable receptacle for the seed; thus all that is needed after the operation is to broadcast the seed, most of which falls in the tracks made by the cart wheel, and to harrow across once or twice. The method was by no means rarely seen when horses provided the power on farms; the tractor-drawn plough with furrow press behind is the modern equivalent.

The operations might then be as one of the following.

Plough	October 10th–15th
Light harrow twice along furrows	November 1st
Light harrow across furrows Drill Seeds harrow	November 3rd
Seeds harrow (across)	November 4th

Or

Plough and ring roll or furrow press	October 20th–25th
Light harrow Drill Seeds harrow	November 1st
Seeds harrow (across)	November 2nd

Or

Plough and furrow press	October 20th–25th
Broadcast seed Harrow twice across furrow	October 30th

(5) AFTER POTATOES

Potatoes are grown chiefly on the lighter soils, and very commonly on fens and silts. On the latter they are often followed by sugar beet or a market-garden crop, but in many instances wheat occupies the field next. Potatoes leave the land in a rich and clean condition, and main-crop varieties are lifted shortly before the usual date of drilling wheat; the final operation in harvesting the potatoes is to harrow, to expose covered tubers and to collect the haulms for burning, and this leaves the field in a very 'tilthy' condition. There are three possible methods of procedure:

Drill Seeds harrow (twice)	or	Cultivate Drill Seeds harrow (twice)	or	Plough Light harrow Drill Seeds harrow (twice)

In a dry autumn the land may be sufficiently loose, after the potatoes have been harvested, for the drill coulters to penetrate, and the first method is cheap and satisfactory. But usually there is sufficient rainfall to make the land too firm as a result of the carting during lifting operations, and the cultivator is put across the field once or twice to loosen the surface and let in the coulters. This is the method probably most commonly adopted on fen and silt soils, which do not tend to set into a solid condition during the following winter. Where, however, the land contains any considerable proportion of clay there is always this tendency, and there is much to be said for the third method, because the plough will thoroughly loosen the soil, and will bring to the surface a certain amount of clod. On this type of soil the weather of the season will determine which procedure is best. Occasionally a summer drought is followed by a dry September and October, and then the potatoes leave the field in an extremely loose and dusty condition; in such circumstances the cultivator will have no effect, and even the plough will merely move the soil sideways, because the loose surface will not hold it sufficiently for it to cut down into the solid ground beneath. When the soil is a little wetter the loosening effect of the cultivator may be all that is needed to get the drill coulters into the ground, though the field may set into a very solid condition during the winter. In a wet October, however, it is only the plough which will loosen the soil satisfactorily; in an extremely wet season the carting may have puddled the soil and it may plough up in a very rough condition, necessitating the use of the drag harrow to break it down again.

Experiments have been conducted to compare the efficiency of these methods, on a soil which contained quite an appreciable amount of clay. The results showed that the relative merits of the methods depended on the wetness of the season, but that in general the plough was to be recommended for that soil type. Unfortunately ploughing for wheat after potatoes favours lodging in the wheat crop—an experimental finding quite in keeping with the views of practical men—and consequently is not advisable (and in fact is rarely done) on rich soils; the ploughing may increase lodging because it gives a heavier crop, or because it leads to deeper sowing. In some cases a potato crop is weedy and a number of weeds may even survive the lifting operations; usually the chief weeds are annuals, and then it may be advanced as a strong argument for the plough that it will kill the weeds by burying them.

(6) AFTER CARTED ROOTS

Wheat not infrequently follows a crop of roots which are carted, in contradistinction to roots which are folded off on the land; the root crop concerned may be mangolds, or the earliest cleared fields of sugar beet. In this case the land should be clean and in fairly good heart, and, having been well worked during the previous year, in good physical condition; the only difficulty involved in this case lies, in fact, in the matter of time, because the field is rarely available until well into November. It should, therefore, be ploughed as soon as possible, the ploughing being kept up to the clearing, so that the former operation may be completed within a day or two of finishing the carting. The plough furrows are best left for a week or so, in order that the newly exposed surfaces may dry, and then the seedbed should come readily, with one or two strokes of the harrow. Generally the carting will have cut up and puddled the land badly on a fan-shaped area radiating from the gateway through which the roots have been taken; this area often shows up badly in the succeeding crop and unless the weather continues wet, some extra working on it is desirable. Thus the operations might be:

Carting roots	Finished November 18th
Plough	Finished November 20th
Drag harrow Light harrow Drill Seeds harrow	November 30th
Seeds harrow (across)	December 1st

Very occasionally wheat is ploughed in when it follows carted roots. The procedure is to broadcast the seed each day on the area that can be ploughed that day, and then to cover it with a shallow furrow. The method has little to recommend it, except as saving further labour and as a means of catching up with the calendar; the crop is rarely a good one, principally because the seed is usually covered too deeply. An alternative method, occasionally employed when it is very late in the season, is to drill and harrow each day's ploughing as soon as it is completed; in this way every opportunity of getting the seed into the ground is taken as long as the weather remains reasonably good, and usually there is no visible irregularity in ripeness of the crop at the time it is harvested.

II. SPRING-SOWN CORN

The principal crops in this category are barley and oats, though wheat and beans are sown in spring fairly commonly, and peas is another crop that may be included. All these crops need approximately the same conditions, but in practice rather more working is given to a seedbed for barley than for the others; this is partly because barley is a weak, shallow-rooted plant, and thorough tillage is believed to give a fuller developed root system, and partly because a good malting sample must be an even sample, and for that a good seedbed is required so that all the plants may come along together.

For spring corn the seedbed should be finer than for autumn corn, because the hardest frosts of winter are usually past by the time of drilling. Where spring corn follows another grain crop the field will be unoccupied for a long time, and the opportunity should be taken to kill weeds during the autumn by stubble cleaning. If drilling occurs very early in spring there will be little chance for seedlings of spring-germinating weeds to be killed, but when it is late they may appear in vast numbers, and thorough harrowing before drilling may account for a multitude of them. The actual date of drilling will be determined by the weather. The general case is that the land is wet throughout the winter, and drilling is done as soon as it is dry enough; this may be as early as February, or may be as unfortunately late as April. It is, of course, the heavy soils which are particularly subject to the weather; on some very light land it is possible to drill on almost any selected date, irrespective of rainfall in the preceding weeks.

For all spring-sown crops it is extremely desirable that the land should be ploughed early, so that the upturned furrows may lie through the winter to receive the maximum effect of frost; except after roots for folding or grassland, the land is available for ploughing by some date between the corn harvest and the end of November. It should be ploughed as soon as the autumn rush of work is over; the aim should be to complete this 'winter ploughing' on a farm by Christmas, but where the acreage for spring sowing is large or when the season is continuously wet or the ground frozen hard, this cannot be achieved.

Then work gets behind schedule and has to be pushed forward vigorously as soon as the weather permits, which may not be until March or even April; in such circumstances operations must be concentrated into a short time, the urgency of sowing outweighing the importance of weathering. Where work is well ahead and a field has been ploughed in the autumn the question arises as to whether it should be ploughed again in early spring. Against this procedure is the extra work involved and also the danger that the winter mould may be buried and unweathered soil brought to the top to give, in the absence of further frost, a surface deficient in crumb. In general, ploughing twice produces the best seedbed with a crumby tilth throughout the working depth; if patience is exercised and the second ploughing deferred until the bottom of the original furrow is partially dry, the friction on the face of the mouldboard will give copious crumb to make a really mellow seedbed. When labour was cheap and horses plentiful the second ploughing was common but it became rare as wage-rates increased. In recent years the trend has been reversed since with tractors the second ploughing can be done quickly and at very little real cost. Some farmers would assert that spring ploughing is never wasted, with the one proviso that it should not be too late as then valuable moisture may be lost from the soil.

(I) AFTER A CEREAL

This is the case mentioned where a good opportunity is offered for stubble cleaning. The surface should be broken, by broad-sharing or by shallow ploughing followed by drag harrowing, immediately after harvest, and left in that condition until the busy autumn period has passed. By late November or December it should be possible to spare the labour to plough the field to the normal depth, in which condition it should be left to receive the benefits of the winter frosts. On many soils little further working will be required. Just before drilling it will be necessary to break down the furrow slices but, if anything approaching the normal amount of frost has been received, that only requires one or two strokes with a light harrow. On heavy land and after a wet, warm winter more may be necessary. The furrow slices may then still be tough, and need a cultivator to break them down. The procedure then would be to cultivate twice and leave for a few days; this cultivation will 'build up the seedbed from the bottom', that is, it will let some of the fine crumb fall down to the bottom of the worked layer, and will bring the yet unweathered clods to the surface. After a few days' drying these should harrow down easily, but until the harrowing is done the surface is rather in an open condition and subject to drying; if, therefore, the season is getting on, and the end of March approaching, many farmers like to push on with the work at this stage, harrowing and drilling within a day or two of breaking the winter furrows. Thus on light soils, or after a cold winter, the procedure might be:

Stubble clean	September
Winter plough	December
Light harrow twice	March 1st
Drill Seeds harrow	March 2nd
Seeds harrow (across)	March 3rd

and on heavier soil, or after a wet winter:

Stubble clean	September
Winter plough	December
Cultivate twice	March 20th
Light harrow twice Drill Seeds harrow	March 27th
Seeds harrow (across)	March 28th

Where two ploughings are given the second would be in the latter part of February on light soil; on heavier land it would probably be into March before the second ploughing was possible and it would take the place of cultivating, the light harrows being used after the new furrows had been allowed a few days to dry.

The reader is referred to p. 107 for the case of heavy land infested with such weeds as slender foxtail. If the autumn working there suggested has been given, it should be completed very early in the new year and the subsequent procedure would be as above, except that there would be little difficulty in breaking down the winter furrows.

(2) AFTER A LEY

This case occurs most frequently in the north and west of the country, oats being the corn crop usually concerned; the advantage of being able to graze the ley well into the autumn is a considerable one. The field should be ploughed (using skim coulters) if possible before Christmas, the furrows being well set up so that they may receive the maximum effect of the winter frost. From February to April, as the weather allows, sowing is carried out, the seed being either broadcast or drilled. In the former case it is sown on the plough furrows between the crests of which it collects, and then the field is harrowed two or three times, whilst in the latter the field is first harrowed two or three times and then drilled. It is important to avoid turning over the furrow slices, and consequently the first harrowing, and possibly the second, should be along the furrows. The procedure is, then:

Plough	December
Broadcast Light harrow along furrows twice	March 1st
Light harrow across furrows	March 2nd

Or

Plough	December
Light harrow along furrows twice	March 1st
Light harrow across furrows Drill Seeds harrow	March 2nd
Seeds harrow (across)	March 3rd

Good crops are often obtained by either of these methods, despite the danger of the hollow condition at the bottom of the seedbed, noted above in the case of wheat after seeds. In the present case, the furrows settle considerably during the winter, and the repeated harrowings help to make the bottom of the seedbed firm.

(3) AFTER FOLDED ROOTS

In the more southerly parts of the country barley is the crop which normally follows folded roots, but farther north it may be displaced by oats. The case will, of course, only arise on the lighter types of soil, because it is only on them that winter folding is possible. Having been worked well in preparation for the root crop, and having been hoed several times during the growth of the roots, the land should be in good physical condition; but the surface inch or two will be very much trodden, and may be in a very puddled and hard condition, if the weather was wet whilst the livestock were on the land. It is the surface layer which is the only cause of trouble in this case, and though it is possible just to plough it under, and do nothing to ameliorate it, successful growers of high-quality barley take pains to break it up, and avoid the hollow bottom which it is apt to give to the seedbed. The magnitude of the difficulty depends on the time available and on the weather encountered. The classical case was that of barley following folding by arable sheep; these arable flocks created the stiffest problems because they spent all their lives on arable ground which meant that they were still folding some land intended for barley in March and even into April. Such flocks are now rare but root folding is still common. In the north lambs born the previous spring are often fattened on roots during autumn and winter, though the current butchers' demand for small lambs is rapidly reducing the number of cases. The most efficient way of using sugar-beet tops is to run sheep or cattle over the field (either in folds or at large) but this should not lead to any serious cultivation trouble; the intensity of stocking (and hence puddling) is low and the tops are usually finished by early February. A development of the last decade is the folding of cows on kale behind an electric fence; in general this finishes by the end of January but if the crop is a heavy one and the winter mild and open it may continue well into March.

When the livestock are off the field early in the winter the trodden surface layer can be weathered but when they are off late there is no time for this and the farmer may have to be content with a seedbed far below the best standard. The two cases will be considered separately.

(i) *Roots folded early*

The field should be ploughed as soon as possible, which may be before Christmas; the depth of ploughing should not be very great, say 4–5 in., because it is inadvisable to bury the trodden layer too deeply. If the furrows are well set up, crests will weather down well and any hollowness below will be beneficial rather than harmful at this stage, as assisting drainage. The question then arises as to whether or not the field should be ploughed again. This has already been discussed in a general way (p. 159) and in the present case there is the special point that a second ploughing will return some of the trodden soil to the surface, the only place where it can receive the weathering required to make it crumble down. Where two ploughings are given the second is carried out, probably, in February, and at least a fortnight should be allowed for weathering subsequently. Before drilling, two strokes with the light harrow should be sufficient to produce a good seedbed, but if the land is dry a roll is often used too. If there is no second ploughing it is very desirable to use the cultivator, to break up the trodden layer below, to bring much of it to the top, and to let down some of the fine top soil. As before, this is building up the seedbed from the bottom, and the operation may well be repeated; a day or two for the clods to dry is then desirable, after which they will crumble down readily under a harrow, a roll being introduced if conditions are sufficiently dry. Thus, after early folded roots the operations might be:

Plough	December
Plough	February 10th–20th
Roll	March 5th
Light harrow }	
Drill }	March 6th
Seeds harrow }	
Seeds harrow (across)	March 7th

Or

Plough	December
Cultivate twice	March 1st
Light harrow	March 5th
Roll }	
Light harrow }	March 6th
Drill }	
Seeds harrow }	
Seeds harrow (across)	March 7th

(ii) *Roots folded late*

In this case there is still the difficulty occasioned by the trodden surface of the ground, but little or no time to deal with it, because the livestock will be on the root crop right to the end of the winter; the golden hoof has a high reputation as a preparation of very light land for barley, but there is little to be said for it when it treads the land till well past the right date for drilling the barley. The usual method of preparing the seedbed is to start with the plough, being careful to keep the work right up to the folding, so that the field may be finished within a day or two of the completion of folding. A few days should then intervene, so that the newly exposed surfaces may dry, and then the seedbed is completed with roll and light harrows, the cultivator not being used because it is better to keep the 'unkind' layer underneath; unless the weather is wet rolling should certainly be carried out, in an effort to crush the trodden layer below, and this is probably best achieved by rolling the furrows before they are harrowed. In some cases it may be better to break the trodden layer before ploughing it under. If the livestock have been on the field in very wet weather, they will have puddled the surface badly, and if the weather then becomes dry, that surface may set very hard, even on light land. It is good practice in such circumstances to cut up the hard surface with a disk harrow (with the edges of the disks set at a good angle to the line of draught), or break it with a rigid tined cultivator fitted with narrow points. After this treatment the layer will not be in a nice 'crumby' condition, but it will be disintegrated so that on ploughing there will not be large spaces beneath the furrows. After the plough, working of the seedbed is again completed with harrows and roll, the latter being very desirable if the weather is sufficiently dry. Thus after late folded roots the procedure might be one of the following:

Plough	Completed April 1st
Roll	April 4th
Light harrow twice Drill Seeds harrow	April 5th
Seeds harrow (across)	April 6th

Or

Disk harrow	Completed April 1st
Plough	Completed April 2nd
Roll	April 4th
Light harrow Drill Seeds harrow }	April 5th
Seeds harrow (across)	April 6th

If the folding is continued very late the preparation of the following seedbed may be completed in even shorter time than shown above; in extreme cases many farmers consider it better to abandon the idea of taking a corn crop, and to substitute for it such a crop as mustard to be cut for seed.

(4) AFTER CARTED ROOTS

This is a very simple case. It should be possible to complete the ploughing of the field by Christmas or soon after, and it is then left to weather till it is fit to drill. One, two, or three strokes with a light harrow may be needed, according to the heaviness of the soil and the amount of weathering it has experienced during the winter. If the furrows lie in a 'sad' condition after the winter a drag harrow, or even a cultivator, may be necessary to break them and loosen the field, but generally light harrows are sufficient. The light-land farmer has a great faith in the efficacy of rolling, and in a dry early spring would often use that implement as well as harrows.

III. ROOT CROPS

Root crops have seeds that are very small compared to those of corn crops; it is true that the so-called 'seed' of sugar beet and mangolds is large, but that is a conglomeration of fruits and the seeds within it are very small indeed. It follows that the sowing of root crops must be very shallow, and this necessitates a particularly fine seedbed. Root crops are sown in spring and summer, and at that time of the year there is a danger that the surface inch of soil, in which the seed is put, may dry out; success is therefore highly dependent on moisture, and since in the drier districts it is unsafe to rely on subsequent rainfall, the greatest care must be exercised that the seed is put into moist soil, even if that requires drilling slightly deeper than is normally desirable. In dry districts it is generally best to follow the drill with a roll, and leave the surface rolled down, so that the surface inch of soil may be pressed tightly around the seed; if the ring roll be used the surface will not be very liable to pan, and it will be broken by row-crop work as soon as the plant rows are at all visible. In wet districts, however, it is usually better to follow the drill with a harrow, as if a wet surface is rolled it may set very hard, and be beaten harder by subsequent rain; in that case a very bad pan may be formed on the top of the land, and instances have occurred where the pan is so hard that the young plants cannot force their way through. As regards cleanliness, the field should be clean of perennial weeds, these being dragged out and burnt (see p. 109) before drilling, if present in any quantity. Annuals will be disposed of by the hoeings the crop will receive, but it is better to kill as many as possible before drilling, so that the young plants may not have to suffer very severe competition as they are coming up. It is therefore good practice to reduce the winter furrows some three weeks or a month before drilling, to give time for the germination of annuals, which can be killed by the final cultivations in preparing the seedbed; this is particularly desirable where mangolds are being sown, as that crop germinates very slowly, and the field may be green with annual weeds before the rows are sufficiently visible for hoeing to be commenced.

Root crops normally follow corn crops, so that there is plenty of time to prepare the seedbed. The season of planting of root crops is a long one, and may extend from March to late July; it will be best, therefore, to consider early- and late-sown roots separately.

(1) EARLY-SOWN ROOTS

Included in this category are sugar beet, mangolds, kale, kohl-rabi and possibly swedes; these crops may all be sown between mid-March and mid-May.

After a corn crop it may be assumed that there will be many

seeds of annual weeds on the field, and, since the land is to be unoccupied during the winter, an admirable chance is presented for stubble cleaning; the surface having been broken, the field is left until, probably, November or December when the pressure of other work permits the ploughing of it. Farmyard manure is normally applied in this case and is carted out in the interval, and though this is best done shortly before ploughing, the better carting conditions, or need to empty yards, often determines that it shall be done in September; field heaps are not usually spread until just before ploughing. The winter ploughing should be to the full depth of worked soil and can often with advantage be accompanied by subsoiling; this operation is particularly desirable, on fields with any tendency to pan in the subsoil, before growing a crop of sugar beet. If such an implement is part of the equipment of the farm a balance, or turn-wrest, plough, is preferable to a common plough in this case, because the former leaves no open furrow; these open furrows will, of course, be filled in by the considerable amount of spring working which will be necessary, but they persist in outline, and very commonly their position can be distinguished throughout the growth of the crop by the thinner plant resulting from the fact that the coulters were inclined to ride over the surface of the hollows.

If the field can be ploughed again in February or early March and subsequently allowed a further month's weathering there will be little difficulty in getting a good seedbed, but when farmyard manure has been applied this procedure is precluded since it will bring the manure to the surface again; this could be avoided by spreading the manure just before the second ploughing, a toilsome business on land already ploughed and quite impracticable in a wet winter. Normally, the field is left from the time of the first ploughing until some date in March, or early April, determined by the weather. When it is sufficiently dry, the furrow slices are broken up, two strokes of the drag harrow being one good method, but if the slices are lying in a 'sad' condition the harrow may merely ride over the top; in that case a cultivator must be introduced, despite the fact that it may bring some of the farmyard manure to the surface to clog drill coulters and hoes later on. If, on the other hand, the winter has been dry and cold, the furrow slices may lie very loosely, and then a roll directly on them will do good work. The furrows being broken, the field should be left until a day or two before drilling, so that weed seedlings may appear. This interval is well worth while, but if very wet weather is encountered the field may settle down into a sticky and intractable condition; then it will be best to cultivate it again and leave for a few more days, when it should be sufficiently dry to proceed. Normally no further deep working is necessary, and all that need be done before drilling is to prepare a fine surface tilth. When many weeds have appeared it is best to undercut all the surface, for which purpose a light broadshare is an admirable implement; this tool is also surprisingly efficient in producing mould. The usual method of obtaining the fine surface is to alternate the ring roll and light harrow, and if suitable weather be awaited, twice across the field with each of these should produce good conditions for drilling; in order to get even depth of sowing it is best to drill on a rolled surface and, as indicated above, in dry districts it is advisable to roll, rather than harrow, in the seed. Thus a typical series of operations would be:

Stubble clean	September
Cart farmyard manure	September
Spread farmyard manure	November 20th onwards
Plough and subsoil	November 25th onwards
Drag harrow twice	March 25th
Light harrow Ring roll	April 15th
Light harrow Ring roll Drill Ring roll	April 16th

As in all cases the above sequence must only be regarded as typical, variations in conditions of soil and weather often necessitating considerable departures from it. One very good modification is to do all the preparatory work up to the point of drilling and then to leave the field alone for a week so that weed seeds in the surface inch may chit. The lightest harrow or broadshare blades set to run no more than an inch deep will then kill multitudes of weeds to give the crop a clean start. The important point is that this final pre-drilling cultivation must on no account be deeper than an inch, as otherwise it will bring nearer to the surface weed seeds previously lying too deep to germinate. The procedure is, of course, much cheaper than pre-emergent spraying and it may be equally effective; on the other hand, drenching rain in the waiting week may postpone drilling seriously.

On heavy land it has been recommended that the field should be ridged up, as for potatoes, in autumn, so that the ridges may be weathered right through, to provide amply the mould that is often difficult to get on such land; in some districts the seed is drilled in spring on the ridges, but they may be harrowed flat before drilling. Very successful crops have been grown by this method, but it is not adopted very commonly. On light land there should be no great difficulty in getting the required condition of tilth, but moisture is often seriously deficient. On very light sands a method commonly adopted with roots is to plough and roll immediately before drilling—in other words to bring up from below some moist soil in which to drill the seed. It is easy to argue against the practice because it means putting dry soil beneath and bringing up soil containing valuable moisture, and exposing it to evaporation. But, after all, the main thing is to get a crop established, and this method is the only one which succeeds on very light land in a dry spring; although valuable moisture is lost from the seedbed, on sandy land the success of the crop will always be highly dependent on adequate rainfall subsequent to drilling.

(2) LATE-SOWN ROOTS

The crops to consider in this group are turnips, swedes and kale, which may be sown from mid-May to late July; the soil is usually on the light side of medium.

As before, a good opportunity is provided for stubble cleaning, and the field should be ploughed up before Christmas, farmyard manure being applied before the ploughing, if it is to be applied at all. It will not be necessary to start the spring working of the field so early as in the above case, but the land should be moved in late April, or early May, as it will be getting quite green with weeds by that time. From then till the date of drilling the field should be worked several times, so that successive 'crops' of weeds may be grown and killed. In the early stages the cultivator may be employed because, though it will loosen up the soil, subsequent working will make the soil firm again. The drag harrow, and later, the light harrow, may be used, whilst the roll will probably be introduced before drilling. There should be no difficulty, with the long time available, in producing the fine seedbed required, but sometimes the soil gets into a 'nobbly' condition, with numerous small clods of about the size of billiard balls; if it is harrowed, and particularly if it is ring rolled, when in a condition intermediate between wet and dry these should crumble down readily, but in a droughty summer this half-dry condition may never arise, and then a tilth may have to be forced with the disk harrow. Clearly the actual sequence will be very variable from case to case, but the following would be typical:

Stubble clean	September
Plough	December
Cultivate twice	April 30th
Drag harrow	May 15th
Drag harrow	May 30th
Ring roll	June 10th
Drag harrow	June 11th
Light harrow Roll	June 20th
Drill Roll or seeds harrow	June 21st

This list would have been considered ridiculously inadequate by light-land farmers of the last century, who gave almost limitless cultivations in such a case; many would plough the field 6 or 7 times, and the classical instance was provided by a farm in Sussex on which, in 1870, a field was traversed 47 times in preparation for swedes. It is not proposed to enter again into the arguments advanced in justification for such labour. It is doubtful if the extra cleaning effect obtained was very great, and certainly the expense would preclude any present-day attempt at emulation; furthermore, so much moving of the land must have resulted in serious loss of moisture.

Where much perennial weed is present it should be worked out of the soil and burnt during April, the subsequent treatment being similar to the above. On very light sandy soil the method of ploughing and rolling immediately before drilling is sometimes adopted in this case. Another modification is that caused by the growing of a catch crop in the long interval between the previous corn crop and the roots. In that case the field is ploughed as soon as the catch crop is all consumed, and the working outlined above is concentrated into a smaller space of time.

In the wetter districts of the country where roots are generally grown on the ridge the only difference in the preparatory cultivations is that the field is ridged just before drilling. Normally the double-mould board plough is used for this, but it suffers from the objection that if many little clods are present they tend to be collected in the middle of the ridges, which are thereby rendered hollow, and liable to drying. This can be avoided by ridging the field up with an ordinary plough, turning a small furrow one way, and another, larger one, on to it from the opposite side; then the clods will tend to roll down outside the ridge, but clearly the speed of working is reduced by half.

IV. POTATOES

The preparation of the land for potatoes merits separate consideration, because the seedbed required is rather different from that necessary for a root crop. Potatoes commonly follow corn, and though some clods are not highly objectionable too many will give a poor hollow ridge, whilst sufficiency of mould is required to form regular ridges.

The autumnal work is the same as that suggested for early-sown roots. Stubble cleaning is desirable and farmyard manure should be ploughed in before Christmas; there is no particular point in using a balance plough, but subsoiling, where indicated by the presence of a pan, is desirable, whilst very deep ploughing is often practised on good potato land, like the silts of Lincolnshire. During March the land is broken down from the furrow slice by one or two strokes of the cultivator or drag harrow, but there is then not usually time to leave the land so that annual weeds may grow. Commonly ridging up will be started forthwith. This is done with the ordinary double-mould plough or by the three-row ridger; for the success of future work it is important that the ridges should be accurately parallel, and either the field should be marked out before, by taking an empty drill over it with the coulters scratching marks, or a marker should be used on the ridger. The vertical distance from the top of the ridges to the bottom of the furrows is about 6–9 in., and the ridges are 27 or 28 in. apart. Sometimes the ridges are split once or even twice, before the potatoes are planted; this will certainly give more mould, because the double-mould board plough is surprisingly effective in pulverising the soil, but each time ridges are split dry soil is put in the centre of the new ridges, and moist soil on the outside, so the repeated splitting may dry the soil to a disastrous degree. Normally the next operation after ridging is to apply the chemical fertilisers (and, sometimes, the farmyard manure) and then the tubers are planted. The work is completed by splitting the ridges over the tubers, and in this if horses are used one of the two horses must be persuaded to walk on the top of the ridge, because otherwise he will squash and displace the tubers not yet covered. Where all the work is by tractor-power, splitting the ridges provides a real difficulty, which can only be solved by a tool-bar mounted in front of, or underslung on, the tractor; some farmers, otherwise fully mechanised, keep a couple of horses for work in potatoes and particularly for drawing out and splitting the ridges. More variation of method is to be found in working land for potatoes than for any other crop. Where farmyard manure is not available, or where it is to be applied in the ridges, two or three ploughings may be given before ridging up; another possibility is to plough in the farmyard manure with a shallow furrow in autumn and to plough more deeply in spring. Some people even cover the potato setts with an ordinary plough and do not ridge the land until the potatoes are well up; there are districts where this is regarded as good farming but the expert potato-grower shudders at such slipshod methods.

The following list is, then, only an example and many variants are possible:

Stubble cleaning	September
Cart farmyard manure	September
Spread farmyard manure	December 1st onwards
Deep plough	December 3rd onwards
Cultivate twice	March 15th
Ridge up	April 1st onwards
Spread artificials	April 10th
Plant Split ridges	April 11th onwards

Potato planters which open the ridges, drop the fertiliser and setts and cover in at one operation are becoming common; they solve the difficulty of splitting ridges and avoid loss of moisture.

V. LEYS AND GRASSES

The seeds to be sown in this case, particularly those of the clovers, are very minute. They need, therefore, to be sown in very fine crumb, with no more than a powdering of earth over them. They will in fact germinate well on the surface of the land, if sufficient rain follows sowing, and in wet districts excellent 'plants' are regularly obtained by broadcasting the seed; but in dry districts failure frequently follows broadcasting, and farmers prefer to drill the seed into the ground, and thus on to moist soil. Short leys are always sown under a corn crop, but permanent grass, and, in some cases, long leys, are sometimes sown on bare ground; if under a nurse crop, sowing takes place when conditions are suitable any time from March to May, and sowing on bare ground may be at the same time, or at the end of summer. On light land, spring-sown barley is commonly used as a nurse crop, whilst on heavy land an autumn-sown crop may be better; all of the three common cereals are used as nurse crops quite extensively.

(1) UNDER BARLEY ON LIGHT LAND

The advantage of sowing under barley on light land is that the soil has been thoroughly worked to a good corn seedbed in February or March, and little more is required to produce the conditions necessary for the small seeds. These may be sown immediately after the barley, but this means that the little clods on top of the barley seedbed will have no time to weather down to crumb, whilst the seeds may, in a wet season, grow so luxuriantly as to overtop the barley; if not then, the seeds should be sown a month or so later, when the barley is established, because coulters or harrow tines should not be put through barley just as it is coming up, lest much of the barley be killed. Any clod left on the top of the barley seedbed should crumble down very readily under a light harrow or roll, and alternative procedures would be:

Seeds harrow	April 20th	or	Seeds harrow	April 20th
Ring roll	April 21st		Ring roll }	
Broadcast }	April 21st		Drill }	April 21st
Seeds harrow }			Ring roll }	

(2) UNDER WHEAT ON HEAVY LAND

The advantage of sowing small seeds on heavy land under an autumn-sown crop is that the surface clods will have been lying fully exposed to the weather all winter, and will therefore crumble down very readily. The corn has, of course, a considerable start over the seeds, and, in order to reduce the shading effect in the early stages, sheep are sometimes run over the field first; this is very risky, however, and in any but very dry conditions the texture of the surface may be ruined. The seedbed was probably quite rough in the previous autumn, so that, if not clods, there will be little heaps of mellow earth where the clods were. These must be shattered and consequently the field should be harrowed or, if very dry, ring rolled; a harrow has in its favour the fact that it will disturb the weeds. After the seeds have been broadcast or drilled the field will be harrowed to cover them, if the land is at all wet, or rolled if it is sufficiently dry; many farmers like to roll directly after drilling seeds or about a fortnight later, because the wheat plants will have been disturbed somewhat by the drill coulters, and rolling, by pressing the soil on to exposed roots, may save some of the plants from perishing. The procedure may, then, be as follows:

Seeds harrow	April 5th
Ring roll Drill or broadcast Seeds harrow	April 6th
Ring roll	April 20th

(3) ON BARE GROUND

There is little that need be said for this case, the final operations being the same as those given above. The advantage of bare-ground sowing is that the young seeds are relieved of competition from an established corn crop, but nothing will be gained if weeds are allowed to dominate the field. It is important, therefore, to sow on a clean seedbed, and it may be desirable to run the light broadshare over the field before the harrow or roll which precedes sowing.

VI. SEEDBEDS FOLLOWING LONG LEYS OR PERMANENT GRASS

In these cases there is, at the commencement of operations, a fully developed sward on the surface of the land, and this sward must be turned in and is better killed. In some cases the sward is a clovery one and well-grazed, in others the pasture is tufty or contains much watergrass, whilst in yet others the field may be nearly or quite derelict, requiring clearing, draining and liming before it can be brought successfully into cultivation. In all cases there is danger of wireworm and wherever possible an advisory officer should be asked to sample the soil for counts to estimate the number of wireworm present per acre; experience shows that the incidence of the pest is very variable. The click beetle lays its eggs in tufts of grass and these eggs hatch out to the well-known yellow wireworm which may persist in the soil for as long as 5 years, though the numbers decrease rapidly in the first and second years of cultivation. Wireworm feed on the roots and stems of young plants, generally doing most harm from March to May and in September and October; birds are very fond of them and so numbers are reduced by turning the land over and exposing them.

In the cropping of field which has been in grass some account must be taken of the wireworm threat but the position has been completely changed in recent years by the introduction of seed dressings (see next chapter) which protect cereals from the pest. Commonly it is convenient to take a corn crop first after grass, in some cases two or three consecutive corn crops, and special methods, such as double drilling, very early drilling and high seed-rate and manuring, were used to circumvent the worst ravages of the wireworm; even so, crops frequently failed. Now that cereals can be protected, at very little expense (less than 10*s*. an acre), they are normally the best to grow after grass and, if the wireworm count of the field is high, it is wise to take more than one cereal crop so that the population of the pest may have declined before another type of crop is grown. The wireworm may attack any crop, though mustard and linseed are nearly immune. Beans, peas, kale, cabbage and rape rarely suffer harm and turnips are somewhat resistant; mangolds and sugar beet are very susceptible in their early stages of growth. Wireworm holes in potato setts do not prevent their growing but holed tubers are unsaleable as a vegetable; early potatoes are normally lifted before the new tubers suffer appreciable harm and, where wireworm are present, maincrop potatoes should be lifted as early as possible. The wireworm will also attack grass and clover plants but in a ley there are so many plants per acre that loss from wireworm usually passes unnoticed.

Leatherjackets may also cause serious harm to a crop following grass, but the danger only persists for 1 year; with early diagnosis it may be possible to save the crop by baiting the field with Paris Green and bran.

Grassland can be broken successfully at practically any time of the year. Generally it is best to deal with heavy land during the second half of the summer and prepare it for an autumn-sown crop of wheat, oats or beans. Light land may be treated similarly, but it is unwise to have light land bare for a large part of the summer and, if broken early, it should be sown with mustard in July, the crop being ploughed in during the autumn; on poor, thin, light land it is better still to sow sheep keep (rape and turnips) to be folded in autumn prior to sowing a spring corn crop. The following examples will illustrate the sort of

FIGURE 3

BREAKING OF PASTURE ON HEAVY LAND

1. Original plough furrows; first ploughing mid-July.
2. Steam cultivated; July.
3. 'Close-up' of (2).
4. 'Close-up' fortnight later, after tractor cultivation.
5. Final tilth; October. Working subsequent to 4: disk harrow twice; drag harrow twice plough; light harrow. Season very dry. Tilth rather too fine.
6. Another field; first ploughed late September; disk harrowed twice after wet fortnight. Seed bed good, though some living turf turned up. Rather deficient in crumb.

(Coins are half-crowns; diameter $1\frac{1}{4}$ in.)

cultivation sequences required. The reader should note the prominence given to the disk harrow in what follows; in the breaking of turf this implement comes into its own, not only because its cutting action is ideal for reducing a turf, but also because it has a great effect in rendering the land firm.

(1) POOR GRASS ON HEAVY LAND

This should be broken about midsummer and bastard fallowed. In a hot dry summer the turf will die very readily, even if it is left loose on the surface after cultivating. In normal conditions it is better to start with the plough; a fortnight or three weeks later this should be followed by another ploughing or a cultivation, with the object of killing the turf before reducing the clods by repeated disking. A final ploughing in September should give sufficient mould and the only subsequent working necessary will be a light harrowing, and possibly a rolling if conditions are very dry. Thus the full operations might be:

Plough	July 1st onwards
Heavy cultivator	July 20th
Disk harrow (4 times)	During August
Plough	September 10th onwards
Light harrow Drill Seeds harrow }	October 5th
Seeds harrow (across)	October 6th

An alternative method is what is known as pre-disking, and this has proved very successful, especially where the field is very rough or covered with ant-hills; if the land is acid, small chalk can be applied to the grass field and then pre-disking ensures a good mixture of it with the soil. The first cultivation is disking with the heaviest disks obtainable and this should be repeated several times until the surface is thoroughly cut up. A plough will then bury the broken turf and two further ploughings should ensure a thorough kill, and produce a good firm seedbed.

(2) GOOD GRASS ON HEAVY LAND

In this case there is not the same necessity to kill all the turf; if some survives all through the first arable year, little harm will be suffered. Thus it is possible to utilise the field as grass all the summer, only taking care to have it grazed down bare at the time of ploughing, which should be in the first half of September. All the green sward must be covered in and it is very desirable to use skim coulters for this ploughing, which need not be a deep one. In moderately wet conditions the furrow slices should be left as turned, but in a dry year a rolling and two diskings along the furrow may well follow immediately. The field may then be left till the time to drill when two or three further diskings will be required; the last two diskings can with advantage be across the original furrow slices, but if this turns up too much green material they should be stopped and working resumed in the direction of ploughing. If no working is done immediately after ploughing, more diskings will be required just before drilling; the operations would then be somewhat as follows:

Plough	September 10th onwards
Disk 4 or 5 times	October 10th–15th
Drill }	October 16th
Light harrow }	October 17th
Light harrow (across)	

(3) POOR GRASS ON LIGHT LAND

If the land is to be broken for autumn seeding, it should be ploughed first in June and reploughed 3 weeks later. In late July or early August it should be disked several times and drilled with mustard. Having ploughed the crop in during October one or two diskings (along the furrows) should suffice to give the required tilth, and the drill should follow the line of the last ploughing.

On land of this description potatoes often succeed remarkably well, though there is the danger of wireworm damage already mentioned. The preparation for that crop should not usually start until about Christmas and then the ploughing should be to the full depth of the soil; deep tilth is essential for potatoes and depth must be obtained right from the start. Having lain in the first furrow for the winter, the land will crumble down readily under disks or even tined harrows, after which it may well be ploughed again. Subsequently one light harrowing will suffice before ridging up. Much cleaning can be done after the potatoes have been planted, by harrowing and ridging up again.

If labour is available and the area concerned is small, double ploughing is an excellent method of breaking grass, particularly on light land. It is usually carried out by two horse-drawn single-furrow ploughs, one following the other in the same track. The first turns a shallow furrow (about 3 in.) and the second takes a further 4 or 5 in., turning it on top of the first furrow slice; thus a thin turf is cut, but it is buried very deeply. The work is slow, but it may be done during winter when horses might otherwise be idle in the stable. With a tractor similar work may be done by fitting a wide skim coulter *behind* a single-furrow plough; the thin turf cut by the skim is thrown to the bottom of the open furrow, to be squashed down by the tractor wheel on the next round. Double ploughing effectively disposes of a turf and subsequent cultivations present no difficulty whether the crop be spring corn, potatoes or roots.

(4) GOOD GRASS ON LIGHT LAND

Light land pasture that has been well grazed can be converted to arable land very easily. A method which has proved very successful is to plough in February, using skim coulters and taking care to cover in all the turf, to give very little working (along the furrows), and then drill barley. Normally, the tilth is obtained by disks, but in their absence light-tined harrows will scratch the upturned furrows sufficiently. A great advantage of the disks is that they will be more effective in making the land firm than tined harrows; firmness is essential to success and a heavy ring roll should follow the plough immediately. Thus the operations might be:

Plough and roll	February 20th onwards
Disk twice	February 27th and 28th
Drill Light harrow	March 1st

The field may be cross harrowed, but not if the operation brings any of the green turf to the surface.

In wet districts oats often follow a long ley and then a common method is to plough during winter, setting the furrows well up, and to broadcast the seed, which falls between the crests of the furrow slices. Two strokes of the light harrow along the furrows and then two across give a good cover to the seed, though the seedbed may be rather 'hollow'.

(5) DIRECT RESEEDING

Poor pasture can be greatly improved by ploughing and seeding down directly to long ley, without taking any intermediate arable crop. As a general practice this is not to be recommended, since it gives no chance of cashing accumulated fertility in arable crops or of killing the weeds and inferior grasses in a worn-out turf, but there are occasions where rapid improvement of the field as grassland is the prime necessity, or where trees or contours debar a field from profitable arable cultivation.

If the field is very poor and the herbage very inferior, it will be well to start in the summer, killing the sward and 'getting the sun into the land', ploughing again in the autumn and leaving in the furrow for the winter; generally, however, the aim will be merely to invert the sward and trust to the better plants to be sown, aided by good grazing and manuring, establishing themselves against the growth from the upturned turf. For this it is early enough if the first ploughing is done before Christmas; it may, indeed, be left until just before reseeding, but then there is some difficulty in getting a fine, firm tilth for the small seeds, whose establishment is therefore somewhat problematical. If the field is ploughed in December, it can be left undisturbed till the land dries in March and should then be disked five or six times and ring rolled before the seeds are sown (preferably drilled) at the end of March or early in April. Often the field is deficient in lime and very poor, and generous doses of lime, together with 5 to 6 cwt. of complete fertiliser, should be worked into the seedbed.

A cold list of typical cultivations conveys but little of the infinite variety and the skill in choosing the right moment for working, which are compounded in the whole art of tillage. In the above descriptions the aim has been to prescribe cultivations which, intelligently timed, should produce good seedbeds, but many things occur which preclude the thorough preparations which good farming demands. It is a counsel of perfection to urge that a farmer should always be ahead of his work, and what is a clay-land farmer to do in a wet autumn? If he rigidly adheres to the rules of tillage he may never get his seed in at all, and often he is right in deciding to drill the field somehow, even though the work done revolts him; that this does not necessarily lead to failure is evidenced by the saying, with regard to wheat —'Sow in a slop, and you'll get a crop'. At other times he may be behind with his work, and he may decide to do no more than plough the field, and to cut the seed into the furrow slice with a disk drill. With wheat and, to a less extent, other corn crops, liberties like this may often be taken with impunity, but in the case of root crops the farmer must take considerable care over the preparation of the seedbeds if, over a run of years, he is to obtain satisfactory plant establishments.

CHAPTER VI

THE CHOICE AND TREATMENT OF SEED

It would be an exaggeration to assert that the success of a crop is determined by the wisdom shown in choosing the seed, because many things, some quite outside the farmer's control, contribute to the final result. Nevertheless, the choice of seed is a very important factor which merits the most careful attention. The majority of farmers recognise this and, although they may not always be well informed on the subject, they do devote some thought to the selection of seed; but there is a considerable minority who fail lamentably in the matter, and very often they suffer a just retribution. Economy on this point may be false in the extreme, but it must not be thought imperative to spend much money in the purchase of seed; what is necessary is a thorough appreciation of the points to be considered before a given sample is finally chosen. These points are as follows:

(*a*) *Suitable variety.* It is not proposed here to enter into the relative merits of the many varieties of the different crops but it is important that a farmer should choose varieties with some care. High yielding propensity must always be an important point but it is by no means all that is required; quality may be as important as yield. With cereals, the standing power of the straw and liability of the grain to grow out during a wet harvest are weighty considerations. With all crops, resistance to diseases and pests may determine a farmer's choice of variety whilst hardiness and time taken to reach maturity must also be taken into account. All these matters, and there are others specific to particular crops, affect a farmer's choice but he must also make sure that the chosen variety is likely to do well on his particular farm; some varieties will give a good account of themselves over a wide range of conditions but some are more specialised in the sense that they only succeed in suitable circumstances of soil and climate. There is no excuse for ignorance of the characteristics of varieties because the National Institute of Agricultural Botany tests old and new ones in different conditions and the conclusions drawn are widely disseminated. Apart from the Institute's own centres, trials by the National Agricultural Advisory Service are widely scattered over the country so that there is plenty of opportunity for seeing a new variety before deciding whether or not to grow it. A word of warning, however, is necessary. Where the same varieties are compared at a number of different centres and experimental procedure is standardised, trustworthy information on the relative yielding powers of the varieties can be obtained, but yields of single or duplicate plots at any one centre may be very misleading; by itself, a centre can only be useful to show variations in habit of growth, standing power of cereals, hardiness or resistance to disease and even these should be checked with the behaviour at other centres. A new variety must be tested for several successive seasons before it can be assessed with a fair measure of certainty and it is only after this has been done that the Institute includes it in the Recommended List. Even then some farmers are slow to change, and cling too faithfully to old and tried favourites. Others go to the opposite extreme and eagerly purchase new and untried varieties; too often they experience the disappointment of the gardener, whose flowers rarely come up to the blooms depicted on the seed packet. The mere fact that a variety is new is no recommendation; before buying seed of a new variety, the price usually being greatly in excess of the commercial price, the farmer should demand real and trustworthy evidence that it is definitely better than the variety which he proposes to displace. It is largely as a result of the Institute's work that the last two decades have seen great improvement in farmers' choice of suitable varieties.

(*b*) *Trueness to variety*. If a farmer has chosen the variety wisely he will want the seed sown to be all of that variety, and not to include some seeds of other varieties; the chosen one is presumably the best for his purpose, and any others included as impurities would therefore be expected to lower the value of the sample to him. Trueness to variety will ensure that all the plants of the crop will mature together, and this is of very particular importance with barley, in which uniformity is a ruling factor in determining malting quality. When it is hoped to sell the produce of the crop for seed, trueness to variety is obviously of importance. It is desirable, however, to see both sides of every question, and there is something to be said from the opposite point of view in the present instance. It is sometimes argued that mixed varieties should be used for the very reason that they do differ among themselves in physiological characters, and in special attributes such as disease resistance; none of the conditions in which the crop will be grown can be predicted really accurately, and it may be hoped that at least one of the ingredients of a mixture will find the circumstances in which it flourishes, or will withstand the attack of a parasite which it may encounter. It is not at all uncommon for mangold varieties to be mixed. There are, of course, many cases where a further step is taken and mixed crops are grown; the main argument advanced in their favour is that the total yield obtained from the field is greater than would be obtained from a pure crop.

With most arable crops the balance of the argument is generally held to be in favour of trueness to variety, and that can normally be assured with little trouble. This was not always the case as is shown by the fact that Le Couteur, farming in Jersey in 1836, had twenty-three different varieties pointed out to him in what he regarded as a pure crop of wheat; he it was who called attention to the need for pure stocks of seed, and by his writings he accomplished much in that direction.

At the present time farmers can obtain pure seed for nearly all their crops. Seeds merchants supply it—they cannot, in fact, afford to sell seed with any appreciable admixture of other than the named variety—and in the case of cereals most stocks in use are fairly true; where a few odd wrong ones appear, and the grain is wanted for seed, it costs little to send men through the crop shortly before harvest to pull off the ears of what are usually known as rogue plants. A big step forward was taken when the Seed Production Branch of the National Institute of Agricultural Botany instituted its cereal Field Approval Scheme. This is voluntary but practically all seeds merchants are members of it; under the scheme growing crops intended for seed are carefully inspected so that when a farmer buys 'field approved' seed he can be sure not only that it is free of weed seeds and disease but also that it is true to variety. The scheme has made great impact on farming in the few years it has been running and now covers crops sufficient to provide seed for over 10 per cent of the cereal acreage of the country. With beans the case is different, because cross-fertilisation occurs and the stocks in the country therefore consist of mixed strains; at intervals selected pure seed is put on the market, but after a few years in general use it becomes mixed again. The seed of root crops is in most cases obtained direct from seeds merchants who take great care to maintain stocks true to variety; they have to lay down rigid conditions for the farmers growing seed for them on contract, and are not invariably successful in avoiding cross-fertilisation, especially with the Brassicae. In the last 25 years important work on strains of grasses and clovers has been done at the Welsh Plant Breeding Station, Aberystwyth. It has been known for more than a century that big differences in habit of growth, longevity and other commercially important characteristics existed within these species. Now the strains have been sorted out and new ones bred, so that farmers can obtain varieties of the commoner grasses and clovers with known performance and can be assured that seed bought from a reputable merchant will be true to variety. In 1956 a National Certification Scheme was introduced for grasses and clovers; as with the Field Approval Scheme for cereals this entails the inspection of the growing seed crop and similar benefits may be expected to accrue from it. Many stocks of sainfoin are very mixed so that the farmer who buys, say, common sainfoin often suffers from a large admixture of the shorter lived, giant strain.

(*c*) *Soundness*. This word is used to embrace many factors which affect the germinative power of the seed; these are, obviously, of the first importance. Good plump seed is desirable, because the seed supplies the food reserve for the plant until the latter is established, and the plumper the seed the greater the amount of reserve. Thin, shrivelled seed is rarely satisfactory and special attention must be given to oats in this respect, because that crop is often cut in an immature stage; Scottish trials have shown that higher yields are obtained where large, rather than small, oats are sown. It is a mistake to cut any crop which is intended for seed until it is fully mature; this is illustrated by the fact that cutting should be deferred until the crop is absolutely dead ripe in the case of barley intended for malting, because a full and rapid germination is required on the malting floor. Colour is another attribute to which considerable attention is paid, again, particularly with malting barley. Discoloration and dullness are signs that the seed has either suffered from the effects of weather in the field or from heating in the stack, and both of these may affect its germination deleteriously; at the same time it is possible to be misled in the matter by varietal characteristics, the best-known case being that of Spratt-Archer barley, which is naturally a rather dull grey colour, but which has a very high reputation for malting purposes. No farmer is likely to buy, for seed, grain which has sprouted, because it has already germinated and died; occasionally, however, home-grown grain, which includes a considerable proportion of sprouted seeds, is drilled, with results which can easily be guessed. Broken seed is also to be avoided, and this calls for care in the case of corn crops which have been harvested in a very dry summer; much grain is then liable to be broken in threshing, and when the corn is intended for seed the setting of the drum must be carefully adjusted to avoid this. Beans are particularly liable to break in the threshing machine (in a dry season a few farmers still thresh beans for seed with the flail), and broken beans are very frequently used for seed. Some farmers, indeed, prefer them to whole beans for seed and they certainly do germinate rather more quickly; but the exposed surfaces collect moulds in the ground and, although laboratory germination may be perfectly satisfactory, there may be a poor establishment of plants in the field. Finally seed should be dry, because damp seed usually has a subnormal germinative power; seed is rarely so wet that the moisture can be seen, but slight dampness can sometimes be detected by pushing the hand into the sack, when it feels cold and it is difficult to force the hand far into the mass of seed.

When a farmer uses seed that he has grown himself he must take all the above points, bearing on the soundness of the sample, into consideration. Generally it will pay to be particular, but in some cases he may decide to use a sample which he knows falls short of the ideal. In that case he should make sure that the germination percentage is reasonably high. For the sum of 2*s*. 6*d*. the National Institute of Agricultural Botany at Cambridge will conduct a germination test on a sample for the farmer's own information; the fee is higher (7*s.* 6*d.* to 25*s.*) when a full test is made and the Institute may be required to stand by the result in a court of law. Half a crown is no very large sum, but even that may be saved if the farmer is prepared to take a little trouble and make a rough germination test for himself. A sample must be taken to represent the bulk and 100 seeds counted out and spread on a piece of blotting paper, which must be left in a warm place and kept moist (a corner of the blotting paper should be in water); after a few days the number which have sprouted will give quite a good estimate of the percentage germination of the bulk. If the percentage is in the neighbourhood of 80, in the case of a cereal, the bulk may reasonably be used for sowing, but the seed rate should be increased slightly above normal. If the germination is less than 50 per cent it is better not to use the bulk for sowing, because tests in the laboratory, or on blotting paper, always flatter the seed to some extent; a low percentage and a slow germination will mean that on the field the proportion of seeds which establish themselves as plants will be less than the number counted as capable of producing a sprout.

When a farmer buys seed from a merchant or another farmer he is safeguarded, to some extent, by the Seeds Act, 1920, and the Regulations issued under it. These require that any seed offered for sale as such must have its percentage germination and purity declared in writing at the time of the sale; the farmer is not precluded from buying and using a sample however bad it may be, but if he elects to do so he has, at least, been warned. There is an 'authorised minimum' percentage germination for each crop, and if a sample of seed falls below this figure its actual germination percentage must be declared; if the figure is equalled or exceeded by the sample a written statement to that effect is sufficient. For the various corn crops the authorised minimum varies from 80 to 90, and for root crops from 70 to 80, except in the cases of mangolds and sugar beet for which it is 60 (percentage of clusters). For perennial rye grass the authorised minimum is 85 and for Italian rye grass it is 80; there is no corresponding figure for other grasses and for clovers, so that the actual percentage germination of a sample must always be declared in those cases. The percentage of hard seeds (see p. 197) must also be declared for samples of seed clovers, medicks, and sainfoin. Although in many cases where unsatisfactory germination occurs no sample has been retained, and the farmer has no means of proving that the seed did not come up to the prescribed standard, the Seeds Act has undoubtedly done much good; the possibility of legal proceedings may be a much more potent deterrent than might be conjectured from the number of cases which actually appear in court.

(*d*) *Age of seed.* This point might have been included in the last, as affecting the germination percentage, but it is scarcely covered by the word 'soundness'. As seed becomes older it loses its germinative power, but the rate at which the loss occurs varies very much between species; onion seed falls off very rapidly, and even when one year old may have largely lost its power to grow, whilst at the other extreme are the oily seeds, which are well known to retain their power of germination for many years. In general, seed is sown in the year following its harvesting, but stocks not used may generally be kept with safety for a further year. A number of tests have been made with wheat, and the general finding is that there is a slight decline over the first 3 years, and then a rapid one over the next 3 years, by which time only a small proportion will still grow. Tales of viable grain having been obtained from the Pyramids must be entirely disbelieved; such grain has maintained its shape surprisingly, but is coffee coloured and quite incapable of germinating. Grain has also been found in the graves of prehistoric men in this country and in grain stores dating back to the earliest times; this grain has become converted into charcoal and, though retaining its original shape, has swollen appreciably, and, of course, has been dead for many centuries. Where over-year seed is grown it is a wise precaution to increase the seed rate slightly, and seed older than 2 years should be avoided. The case of beans is an interesting one. For feeding purposes old beans (1 year old) are better than new beans (the latter cannot be ground as they are too soft) and, during the autumn months, separate prices are always quoted for the two sorts; it is, therefore, common practice to keep a stack of beans for a full year before threshing it. As regards feeding value the position is quite clear, but what is not so clear is the relative merits of old and new beans for seed. Some farmers go to the trouble of threshing out a few new beans for seed every year for sowing in the autumn after they are harvested, although they keep the bulk unthreshed for use as feed in the following autumn; on the other hand, there are farmers who do not usually keep beans for feeding a full year in the stack, but nevertheless leave a small stack unthreshed for a year, in order to have old beans for sowing. There are thus contradictory indications in practice of the relative merits of old and new beans for seed. A series of experiments was conducted at Cambridge to solve the conundrum presented by this contradiction, and the conclusion reached was that age was not itself an important point; new and one-year-old beans are, in general, both satisfactory for seed, and the choice between them should be determined by the conditions when they were harvested. Beans are left in stook for a long time, and in a wet harvest they are apt to lose some of their germinative power. After a wet harvest therefore, old seed will probably be preferable (if the previous harvest was also wet, the beans will have fully matured and have dried in the stack), but after a dry harvest new seed will be just as good and may, indeed, be better. It is difficult to say whether practice varies from district to district (it varies, rather, from farm to farm), but if it could be established that farmers in a dry district tended to use new beans for seed, and those in wetter districts used old beans, practice could be said to be justified in both cases.

(*e*) *Purity of seed.* Absence of weed seeds is clearly an important characteristic of a good sample of seed. The point has been discussed in Chapter III and consequently need only be mentioned here. The Seeds Act and its Regulations cover the question of purity as well as germination. Samples of seed must have the percentage of weed seed present declared at the time of sale. It is made an offence to offer for sale any seed which includes more than 5 per cent by weight of injurious weed seeds, these, for the purpose of the Act, being docks, sorrels, cranebills, wild carrot, Yorkshire fog and soft brome grass. The actual percentage content of these injurious weeds must be declared for samples of grass seeds if it exceeds 2 per cent of the weight of the sample, and for the clover seeds if it exceeds 1 per cent. The presence of dodder must be declared if there is as much as 1 seed in 1 oz. of the seed of wild white clover, in 2 oz. of timothy, alsike and ordinary white clover, or in 4 oz. of red clover, crimson clover, lucerne and flax.

(*f*) *Freedom from disease.* Some fungoid diseases are seed-borne, but the importance of ensuring that they are not carried in a particular sample of seed depends on whether or not it is possible to treat seed against them. Bunt in wheat, and leaf stripe in barley and oats, are seed-borne but, as will be seen later in this chapter, they can be satisfactorily controlled by a simple treatment of the seed; it is not, therefore, necessary to take care that they are absent from a purchased sample, as long as it is the routine of the farm to treat cereal seed. Grain of wheat and barley carrying the fungus responsible for loose smut on the other hand, can only be disinfected of it by a treatment which is difficult under farm conditions and is, in fact, rarely carried out; it is only possible to ensure that grain is free from this disease by viewing the crop which provided the sample, and it may be difficult to find crops of wheat and barley which are free from loose smut. The importance of freedom from disease is greatest in the case of potato-seed tubers, and it will be seen later in this chapter that this is the reason for the large seed-potato trade from Scotland to England.

(*g*) *Change of seed.* This final point presents the difficulty that farming opinion has been almost unanimously in favour of a practice for which no adequate scientific reason can be provided. The subject is still hotly debated and a number of practical men hold to the view that the seed should be changed from one set of conditions to another; thus many heavy-land farmers like to obtain seed which was grown on light land, and vice versa. The belief may have arisen largely over the matter of disease. Thus leaf stripe is very prevalent in oats grown in the south-west of Scotland, and is relatively scarce in Aberdeenshire. For some time there has been a considerable demand in the former district for seed from the latter, but trials have shown that in the south-west of Scotland home-grown seed treated against leaf stripe gives as good a crop as seed brought from Aberdeenshire. Here was a case where change of seed was undoubtedly very beneficial, until a satisfactory control of the particular disease could be achieved; and it is probable that the idea that change itself was good received much support from this and similar causes. One possibility is that the seed is better physiologically when produced in some districts, which therefore become suitable ones from which to obtain it. Thus some farmers in Norfolk change their seed oats every 3 or 4 years and seed wheat every 7 or 8 years, whilst they keep on growing the same stock of barley indefinitely. The conditions may be described as very suitable for barley, moderately suitable for wheat and somewhat unsuitable for oats. In the case of oats, therefore, it may be that the grain is slightly poor physiologically, and this gives a bad start to the next crop, in which a further weakening occurs and so on; thus in 3 or 4 years the stock has accumulated enough physiological weakness to justify a fresh start with seed from a district which suits the crop. This, of course, is no general rule in favour of change of seed, but it might provide part of the explanation of the wide belief in the desirability of change of seed; in the case of some crops it may be that certain districts are unfavourable for the ripening of the seed and hence that seed matured elsewhere is desirable. There is a contrary argument which may be put in the case of a crop to which the conditions are not suited. If a stock of seed be introduced, those plants which are most suited will thrive best and will, therefore, contribute more than their quota to the bulk of seed obtained; thus if the stock be grown for several years there will be a constant natural selection from it of the physiological strains most successful in the given conditions. In support might be quoted the fact that if spring oats are sown in autumn, though mortality is very high, the surviving strains may display a marked increase in winter hardiness over the original bulk. It is possible, therefore, on purely theoretical grounds, to argue against change of seed. It is clearly unwise to be dogmatic in the face of so much practical opinion, but the probabilities are not in favour of any advantages being gained from changing seed, except where that is the only method of obtaining a sample which is well ripened and free from disease.

WHERE FARMERS OBTAIN SEED

It was common for cereal growers (except those who believed fervently that there was virtue in changing seed) to use seed from their own crops but practice has changed perceptibly in recent years. A fuller awareness of the importance of clean, sound seed and of the value of seed treatment (see below) has led to much greater reliance on seeds merchants, even though the cost of seed from them is much above the corresponding market price. The seeds merchant is equipped with machinery to remove practically all weed seeds and to apply seed dressings and now, under the Field Approval Scheme, he gives useful service in keeping stocks free from admixture; he is, in fact, a specialist who is entitled to recompense for his work, though some would argue that his margin is unduly high. Where a new variety is being marketed and a farmer is determined to give it a trial on his farm, he usually buys a small quantity of seed and sows it (thinly, so that it may be multiplied more) on a few acres, to provide him with seed for the following year. This is pure common sense, because the original stock may cost three or more times the current commercial price. With root crops seed is usually obtained direct from a seeds merchant every year; in parts of Scotland, however, it is a common practice to leave a small area of a root crop in one corner of a field, to grow on another year and to provide seed. The seed of sugar beet is normally obtained from the factory for which the crop is being grown, the cost being deducted in the autumn from the payments the farmer receives. The seed of grasses and clovers is usually obtained from one of the larger seed firms, which will make up any mixture to a farmer's prescription. There are, however, some farmers who have a stock of one species (often red clover) with which they are pleased, and who always keep some back to maintain the stock on the farm.

A special word is necessary on the selection of seed-potato tubers, in which two important considerations are size of the setts and district in which they were grown. If large potatoes, containing many eyes, are planted they give rise to bushy plants producing many small tubers; if small potatoes are planted they produce few, but large, tubers and the total yield is about the same as from big seed. From this, very small seed would appear desirable, but if the small ones are riddled from a bulk the sample obtained will contain not only the tubers which are small for physiological reasons, but also those which are small because the plants which produced them were diseased; thus an unconscious selection of unhealthy tubers will be made and this may have disastrous results. In practice, therefore, the medium-sized tubers (generally those passing through a $1\frac{5}{8}$ in. riddle and standing on a $1\frac{1}{4}$ or $1\frac{1}{2}$ in. riddle) are used. The district of origin of the seed is very important because of a group of diseases, known as virus diseases; these are systemic and carried over from one crop to the next in the seed tuber. Virus diseases are very variable in the harm that they do, but in many cases an attacked plant is killed or so dwarfed that it produces few or no saleable tubers, so that considerable trouble must be taken to ensure healthy seed. Crops grown in parts of Scotland (generally speaking the northern parts), Northern Ireland and on some farms in Wales and England are relatively free from virus disease, because the insect vectors responsible for the spread of the disease are scarce in those districts. There has sprung up a very large trade, now under Government supervision, of seed potatoes from Scotland and Ireland to England; English growers use Scotch or Irish seed or 'once-grown' seed, that is, tubers produced, possibly on their own farms, from a crop planted in the previous year with Scotch or Irish seed. Tubers 'twice grown' in England usually produce unsatisfactory crops, whilst those farther away from the original Scotch seed should never be used in the southern counties. It is possible to obtain a stock fairly free from virus in England if it be grown in an exposed situation isolated from other crops of potatoes (a distance of at least 60 yd. is required), and if affected plants be dug up and burnt as soon as they are noticed; this, however, involves much care and uneconomical working, and English farmers choose, rather, to use Scotch or Irish or their own or other farmers' once-grown seed. The amount of seed certified (i.e. Scotch, Irish or that grown on approved English or Welsh farms) used is sufficient to plant about 60 per cent of the English acreage; much of the remainder is planted with once-grown seed, but there is still some land planted with seed farther removed from certification, and farmers thereby suffer very heavy losses in yield. It is still not many years since the so-called 'degeneration' of potato stocks was first shown to be entirely due to infection with virus, and there remains considerable possibility of advance in potato growing through the wider provision of virus-free seed tubers.

SEED TREATMENT

The eminently reasonable idea of treating seed has always appealed to the crop husbandman, because the seed is the crop in miniature, and any benefits of treating a mere sackful or two of seed will be disseminated over a whole field. The earliest references to the subject in the literature describe treatments which were frankly mystical; thus seed was treated with such substances as ground toad or virgin's blood, the treatment in the latter case being attended with horrible rites. In mediaeval times and down to the eighteenth century, the general principle was to manure the seed, by dipping it in what were known as 'prolific steeps'; the literature over a long period abounded with descriptions of how to mix the dungs of the different farm animals, and other materials, in order to produce a steep of amazing efficiency. These old writings are now, of course, regarded humorously, but the idea of manuring the seed, in the then state of knowledge, was quite sensible and well worth trying; in the absence of experimental stations the futility of the procedure was not exposed for a long time, and consequently the practice lingered on for many centuries. Presumably the fact that manuring the seed is futile may be considered established by the very long period of trial to which the practice was subjected; if any room for doubt remained it would be interesting to speculate as to the possibility that auxins, essential for plant growth and present in animal excreta, might provide the explanation.

Though old methods are now thoroughly discredited seed may, and should, be treated with a variety of objects. If potato-seed tubers be left for consideration later, the various modern treatments may be grouped under four heads.

(1) TO ASSIST GERMINATION

The maltster normally kiln-dries barley as soon as it reaches his warehouse; he does this largely because much of it must be left for some time before it is used, but the drying appears to complete the maturation process, and thus to favour rapid and even germination. This would be more important in the wetter districts, and in Eire, where barley growing is well organised, the maltster usually takes all a farmer's barley and dries it, the farmer taking back the requisite amount for sowing. Thus, in some cases, seed may be dried to encourage it to germinate, but more commonly germination may be aided by enabling the moisture in the soil to get to the seed more quickly; the greatest chance in this connection is offered by sugar beet and mangolds, with which the seed is encased in a tough fruit coat.

Sugar beet 'seed' is a cluster of fused fruits so that each can give rise to more than one plant and since the plants from one seed cluster are very close together and often entwined singling is a laborious job. There are therefore two objects in treating the seed: to hasten germination and to break down as many of the seed clusters as possible into single seeded entities. One treatment that was fairly successful but never gained wide acceptance was with concentrated sulphuric acid in which the seed was put for 20 minutes; this was followed by thorough washing under a rapid stream of water and the neutralising of any remaining acid. Experiments showed that the treatment was beneficial, particularly when sowing conditions were dry; it gave quicker and fuller plant establishment and increase in yield of crop. But the treatment requires proper facilities, because the acid froths up copiously and it would be very dangerous to attempt it on a farm. Reduction of the seed clusters mechanically would seem to offer the best solution to the problem. Milling between rollers will reduce the seed clusters and crack them but it may harm germination, especially if the seed is a little green. In America drastic treatments known as decortication, segmenting or shearing have been used. In these the seed is passed between emery wheels or between a steel shear plate and a rapidly rotating carborundum wheel. These treatments certainly reduced the seed clusters and even produced some of the true seed without any fruit coat attached to it; but the action was too drastic and resulted in seriously reduced germination and malformed plants.

The modern treatment is 'rubbing and grading' and this is being used more as experience shows the gains from it. The seed is passed between an emery wheel and a rubber pad and then is graded by sieves so that it is all from $\frac{7}{64}$ to $\frac{11}{64}$ in. in diameter. There is no great loss of seed in processing and a reduction of only about 5 per cent in germination. A 10 per cent saving in singling time is claimed although the percentage of single seeded clusters is only raised from (roughly) the 40 in untreated seed to 55. The weight of seed required per acre is greatly reduced and further advance is being sought by a narrower grading ($\frac{8}{64}$–$\frac{10}{64}$ in.) and by using only the heavier clusters that come into the grade, since these give better plant establishment than the lighter ones.

Treatments for sugar beet seed should be applicable to mangolds which have similar seed clusters but much less experimental work has been done with mangolds. It is an old practice, still sometimes followed, to soak mangold seed in water; the sack of seed is immersed overnight and the next day the seed is spread out to dry, as wet seed will not run in the drill. The object of soaking is to soften the fruit coat and to quicken germination, mangolds being a crop which suffers much from slow and imperfect plant establishment. Experiments have shown that the treatment may be good, bad or indifferent in its effect. In wet seeding conditions soaked seed is neither better nor worse than unsoaked seed; in conditions of moderate moisture, before or after drilling, it may be wholly beneficial as the amount of moisture in the soil may then be sufficient to give full plant establishment; in dry seeding conditions the soaking may prove disastrous, as it may cause the seed to chit and the very young plants, finding too little moisture around them, may die. With such dependence on conditions no general recommendation can be made, but the fact that many successful growers of mangolds in the eastern counties and the midlands have made a practice of soaking the seed suggests that in those districts conditions favour the treatment more often than not; that, in fact, the soil at sowing is rarely so dry that there is a danger of the seed starting to grow as a result of soaking and then not finding sufficient moisture in the ground to continue its life. Water soaking is sometimes carried out with other seeds, particularly with peas by market gardeners, but it is a 'hit or miss' treatment with uncertain results.

Trefoil seed is commonly milled. As threshed, trefoil consists of some true seeds, some unbroken pods and some broken pieces of pod, and is known as trefoil 'cosh'; seedsmen can extract all the seed and remove the pod, thus providing a sample consisting only of the true seed. Farming opinion favours this true seed for sowing in a dry spring, but holds that the cosh is just as good in a wet spring. This is in close conformity with the views expressed above in regard to acid-treated sugar beet seed. Sainfoin when threshed yields all unbroken pods but seedsmen can, with difficulty, mill the pods and extract a clean sample of the true seed; both unmilled and milled seeds are commonly used for sowing, and the writer knows of no experiment conducted to compare them, but by analogy it may be argued that the milled seed would do better in a dry spring, and that there would be no difference between them in a wet spring. Some seeds of the clovers have a water-tight envelope all round them, and when sown will not germinate until this envelope has decayed sufficiently to allow water to enter, a process which may take a year or more; seeds with this delayed germination are called 'hard' seeds. About one-third of the hard seeds in a sample germinate within a year of being sown, and it may be that they are sometimes valuable in providing plants where the earlier established ones have died; there have probably been many cases where the hard seeds of wild white clover have, all unbeknown to the farmer, been the salvation of a long ley or permanent pasture. In general, however, hard seeds must be regarded as less valuable than normal ones, especially for short leys. Clover seed is therefore sometimes put through revolving wire brushes, or scratched by other means, to puncture the envelopes of the hard seeds.

(2) TO DISINFECT AGAINST SEED-BORNE FUNGOID DISEASES

This is by far the most important group of seed treatments, and considerable progress has been made in recent years in this method of controlling those fungoid diseases which are carried from one crop to the next in the seed. The following are the important cases:

(i) *Bunt (also known as covered smut or stinking smut) in wheat.* This disease has caused very serious loss to farmers in the past, but is now, happily, under control. The disease is carried to the crop by spores adhering to the seed and these spores germinate with the grain, the fungus growing in the wheat plant; eventually the places of some or all of the grains of an ear are taken by bunt balls, which are about the size of a grain but consist of a grey crumbly mass of bunt spores. When an infected crop is threshed the beaters of the drum break up the bunt balls and the spores are set free and settle on healthy grain; in a bad case the threshing machine may be enveloped in a black mist of these spores and there is a strong fishy smell. The important point to notice is that the spores are carried to the next crop on the outside of the seed grain, and it is thus possible to kill them by seed treatment; since there are several chemicals which do this it is a reproach for a farmer to suffer from the disease and, in fact, now that seed dressing for wheat is almost universal, attacks of bunt are very rarely seen.

Methods of dressing are continually being improved and more effective and safer chemicals are being discovered. Until about 1925 wet dressings were used, these giving place gradually to dry ones, whilst in future it may be common to use damp ones. The great drawback to wet dressing is that seed must be sown soon after treatment, as otherwise there is a progressive fall in germination percentage. The general practice was to treat one day and sow the next but bad weather might intervene and then, if the delay was more than a day or two, the seed had to be spread out and turned several times to dry it thoroughly. It is an outstanding advantage of dry dressing that treated seed may be kept for a long time without losing its germination percentage; in dry storage conditions it may be kept for a full year though usually the interval between treatment and sowing is a matter of days or weeks. This means that seeds merchants can treat all their seed before sending it out, which is a great convenience to their clients. Dry dressing suffers from a disadvantage compared to wet dressing in that it is much more difficult to ensure that each seed grain receives a light powdering as opposed to being wetted. Some farmers have not appreciated the necessity that each grain should receive a little powder and have even filled the drill with wheat and sprinkled the dust on top in the hope that the jolting of the drill during sowing will give a passable mixing; this is by no means good enough. Neither is it satisfactory to spread the seed on the barn floor, to spread the dust over it, and to turn the heap two or three times with a shovel. In the War of 1939–45 a cheap 'machine' with no moving parts was introduced for this treatment. The seed with the appropriate amount of dust was put into a hopper and the withdrawal of a shutter allowed it to pour over perforated baffles; the contraption was about the size of a household dustbin and gave useful service, but the dusting of the seed was far from perfect and the method was wasteful of the powder. Old butter churns in which the seed and the powder are whirled for 20–30 seconds are satisfactory for small lots but for most cases a continuous working machine is required. Various makes and sizes of these are available, the general arrangement being that the grain passes in a steady stream from a hopper into a revolving drum, the powder being dribbled on to it as it enters the barrel; these machines are engine driven and incorporate a fan to suck out excess powder from the atmosphere within the barrel. Farmers who grow a large acreage of cereals (it will be seen below that the other cereals should also have this treatment) are usually equipped with one of these machines but for others the treatment is generally done by seeds merchants who have large machines and proper protection for the men working them. This last is very important because the powders used are poisonous and it is dangerous to breathe for more than a short time in an atmosphere charged with the dust. For this reason damp dressings are being tried. A wettable powder is used and only a little water is added to form what is known as a 'slurry'; in this method there is no dangerous dust and so little water that drying of the treated seed is unnecessary. As yet the slurry method has been little used in this country.

The original wet dressing used was copper sulphate (bluestone). Treatment with bluestone became fairly common during the second half of the nineteenth century and continued well into the third decade of the present one. The strength of solution was 2½ per cent (2½ lb. of crystals in 10 gallons of water) and up to half a gallon of the solution was required for a bushel of seed. It was either sprinkled over the seed, which was then turned over several times with a shovel, or the seed was immersed in the solution for 30 seconds. After treatment the seed was spread out to dry and only sacked up just before taking it out to the field to drill. Shortly after the War of 1914–18 it was shown that 0·15 per cent formalin (1 pint of 40 per cent formalin in 35 gallons of water) was better than bluestone; the methods of treatment were the same but since formalin is a gas the treated seed had to be piled up and covered with sacks soaked in the solution for 4 hours. A number of experiments proved formalin to be better than bluestone both in control of bunt and as being less likely to cause any loss of germination to the seed; nevertheless formalin has never been used much because shortly after its superiority was established the modern dry dressings became available.

The first chemical used as a dry dressing was copper carbonate but this was rapidly displaced by certain organic mercury compounds. There is a large number of these latter chemicals which might be used and the aim of research is to select one which has the maximum lethal effect on bunt spores without any danger of lowering the germination of the seed and is least likely to harm men and animals. They appear on the market in proprietary forms. An effective chemical which is non-toxic to men and animals may be found, but at present the poisonous nature of organic mercury powders must be kept very much in mind. Seed treated with them must on no account be fed to livestock (except poultry) unless it has been thoroughly washed; sacks which have been used for treated seed must also be washed before use for untreated grain.

The amount of powder required is normally 2 oz. per bushel of seed (there is talk of more concentrated powders of which only ½ oz. is required) and the total cost of the treatment works out to a very few shillings per acre of land sown. For this expense the farmer has the certainty of perfect control of bunt, with no risk of impairing the germination of the seed. Indeed, it has been claimed that the organic mercury compounds have a valuable stimulating effect on the seed of many crops; this claim has never been substantiated and the idea has probably arisen because better germination may follow the killing of some pathogenic organisms. What cannot be denied is that treatment with the powders protects the seed, when it is in the ground, from moulds. The most useful instance of this is provided by peas, which are very apt to rot in the ground when sown early; experiments have shown that dusting with organic mercury dusts has an extremely beneficial effect on peas sown in February, but no effect on those sown later.

(ii) *Leaf stripe in barley and oats. Smuts of rye.* Leaf stripe is a serious disease causing poor plant establishment, a characteristic striping of the leaves and abortive ears. It is carried from one crop to the next in the form of mycelium beneath the paleae, and has been the cause of much loss to farmers. Fortunately it is adequately controlled by organic mercury dusts, used in the same way as described above for wheat. These powders will also control covered smut of barley and oats and loose smut of oats. Rye is subject to a stripe smut and is also susceptible to the bunt of wheat, though the latter is rare on rye; both of these seed-borne diseases are controlled by organic mercury dusts. There is, therefore, one very simple, but extremely important, general rule of crop husbandry—that the seed of all cereals (wheat, barley, oats and rye) should be treated with one of the organic mercury dusts; there is no reservation to make and no exception to what should be an unalterable rule. With wheat the practice is general, though not universal, and bunt is now rarely seen; in recent years intensive propaganda has made seed dusting of the other cereals common, but much untreated seed is still sown and hence it cannot be said that the other diseases mentioned in this section are yet under control. There has been, however, a steady improvement during recent years.

(iii) *Black leg in sugar beet.* This disease can be caused by a seed-borne fungus, but it may also be caused by other fungi living in the soil; it is believed in fact that in some cases no fungi are responsible, but that the symptoms arise through a bad physical condition of the soil. When a plant is attacked, black spots appear on the seedling at ground level soon after the plants emerge; the spots get larger and coalesce, and when the little stem is encircled the plant heels over and dies. The seed-borne fungus is not killed by the sulphuric acid treatment, but can be by treating the seed with organic mercury compounds; the treatment is carried out by seed-producing firms, usually without the farmer knowing anything about it. The treatment is effective against the seedborne fungus, but clearly cannot entirely protect the plant from black leg.

(iv) *Loose smut of wheat and barley.* When the ears of plants attacked by loose smut emerge all the grains (very occasionally only some of them) are displaced by masses of spores; the masses burst as the ears emerge, and the spores are scattered through the crop, where they alight on the healthy ears and enter the young grain at flowering time. Ears which have spread spores through the crop can easily be detected in a field before harvest, as they blacken and stand out as gaunt spectres among the healthy ones. When grain on which spores have settled is threshed it carries the fungus, not externally as spores, but internally in the form of mycelium. Thus it is impossible to control the disease by any of the treatments described above. A method is available, however, which depends on the fact that there is a small range of temperature over which the mycelium will be killed and the germination of the grain not impaired. The method is to soak the grain in cold water for 4 hours and then to put it in water between the temperatures of 125 and 129° F. for 10 minutes, after which it is spread out to dry. The difficulty of this treatment lies in the fact that it is a very narrow range of temperature which will kill the fungus and not the grain. Guesswork is no good over the matter, and farmers are not accustomed to handling thermometers. It will be realised that putting the seed into the water will lower the temperature of the latter considerably, and that much further cooling will occur in a period of 10 minutes; it is necessary, therefore, to have a good supply of hot water available to add, and to keep a careful eye on the thermometer. The treatment is admittedly not an easy one to carry out on a farm and, indeed, is very rarely performed in this country. Loose smut on wheat and barley is fairly common and there appears to be variation between varieties in susceptibility to it. It is possible that entirely resistant varieties will be bred but the likelihood is that hot water treatment will have to be used more in future if the disease is to be adequately controlled. Although difficult on a farm, the treatment can be done effectively and safely with thermostatically controlled tanks and in a few places such equipment is available. The disease is harmful and widespread but it is not a real menace since if its prevalence progresses control can soon be provided.

(v) *Ergot in rye.* Although ergot may attack wheat it is most commonly found in rye, and since that is a cereal of minor importance in this country, treatment against the disease is not common. Spores of the fungus enter the grain at flowering time, and the grain is replaced by a mass of spores called an ergot, which can easily be seen, being bluish black and longer than healthy grain. Some of these fall to the ground where they may infect a crop next year, but many remain on the ear and are found, often broken, in the grain after threshing. The ergots may be separated from the grain by placing the mixture in a 20 per cent solution of common salt, when the ergots rise to the surface and can be skimmed off. This treatment is sometimes carried out on rye grain which is intended for feeding, because ergots are poisonous.

(3) TO PROTECT THE SEED OR YOUNG PLANT

Birds are very adept at finding seed after it has been sown and some crops suffer severely from their depredations; cereals and pulses sown late in the autumn or early in the spring are at greatest risk since there may be few other crops in the neighbourhood in the vulnerable stage. This stage lasts for a considerable time as often the greatest harm is done when the young plants are just coming through. Scarecrows are useless and farmers are no longer permitted to engage small boys at nominal wages for bird scaring; all sorts of noise-making devices have been tried but they all quickly lose their deterrent effect as the birds become used to them. Several substances have been tried as seed dressings to repel birds but success has been very limited. Lime, chalk and charcoal have been dusted over the seed in the hope of discouraging birds but it is a vain hope. Tar has been used and a few farmers have claimed success for it. One pint of tar diluted with one pint of hot water provides sufficient for two bushels of seed; the liquid is sprinkled over the seed which is then turned two or three times so that each seed carries a speck of tar and quicklime is dusted over the heap to dry the grain. Beans and peas must not be treated with tar because of the danger of a little tar sealing the micropyle and so preventing water from entering. There are proprietary repellents on the market but none has been widely used; red lead is reputed to be effective but it is too expensive for farm crops.

Seed dressings for pests other than birds are being developed and a great step forward was made when the insecticidal properties of benzene hexachloride were discovered. When this chemical is applied to the ground it may for several years cause taint in subsequent crops, particularly roots and potatoes, but it was found that it was the gamma isomer which was the effective insecticide and if this is used in the pure form the amount required is very small and no taints occur. Two very useful seed treatments have developed from this knowledge, one to protect cereals from wireworm and one to protect brassicae from the turnip flea beetle.

In regard to wireworm, there are now proprietary cereal seed dressings which are dual purpose; they incorporate organic mercury compounds and gamma-BHC. The method of treatment is exactly as described for control of bunt in wheat and the cost of this double protection is no more than 10*s.* per acre sown. Unless the wireworm population of the field is very high cereals can now be sown safely after long leys and, as mentioned on p. 176, this has had considerable effect on alternate husbandry rotations. Sugar beet can also be protected from wireworm attack by treating the seed with gamma-BHC dressing.

The turnip flea beetle (or 'fly') has in past years been a very serious curse to farmers and gardeners. It attacks the cotyledons of brassicae, sometimes even before they get through the surface of the ground, making distinctive little round holes in them; if the attack is not too severe and growing conditions are good the plants may grow away from the pest and it is generally, but not invariably, true that when the plants have developed broad leaves they are safe from the pest. Ideal conditions for the pest seem to be a bright day in May with a cool east wind; in such circumstances a young crop may be ruined in one afternoon and there is nothing for it but to drill again. The fly may attack at any time from March to July and occasionally even to the end of August and many a farmer has had to drill kale three or four times and then had only a very moderate crop. All sorts of dodges have been adopted to defeat the fly but none of them is more than a palliative, so the discovery that gamma-BHC as a seed dressing gave protection to the young plant through the cotyledon stage was of first-rate importance. Proprietary dressings are available and cheap and they have changed the whole position; the protection they give may break down very occasionally, when all else favours the fly, but generally it is complete.

These new chemicals which will protect not only the seed itself but also, for a period, the plant to which it gives rise are opening up a new and exciting field of plant protection. Other uses will no doubt be found for them. For instance, there is evidence, inconclusive as yet, that wheat sown late in the autumn, the seed having been treated with one of the new dual-purpose dressings, may be protected to some extent from the wheat bulb fly. New chemicals are appearing every year and aldrin and dieldrin are two for which great claims are made. If cereal seed has a moisture content of over 16 per cent at the time of dressing and then has to be kept for any length of time before sowing there may be some loss of germinative power; this danger is reduced if aldrin replaces gamma-BHC as insecticide and there are already rumours of alternative fungicides for the organic mercury ones.

(4) TO ADD SYMBIOTIC BACTERIA TO THE SEED OF LEGUMES

The successful growth of leguminous plants depends on the presence in the soil of the appropriate bacteria, which produce the root nodules. Most of the common legumes have been grown long enough and widely enough in the country for there to be little risk that the symbiotic bacteria may be lacking, but this is not the case with lucerne. Where this is grown in a new district, or on a field which has not carried it previously, failures are by no means uncommon, and experiments have shown that in those cases a successful crop may be grown if the bacteria be added to the soil or to the seed. They can be added to the field by spreading a few loads of soil from an established successful lucerne ley, but in practice this method is laborious and does not always achieve its purpose. The method of adding the bacteria to the seed is very cheap and easy, and is strongly to be recommended in instances where there is any danger that the bacteria may be lacking. Having obtained the culture* (and the accompanying phosphate) all the farmer has to do is to follow the instructions which accompany it. These are to mix the culture with skim milk (not whole milk, as that contains fat which will prevent the treated seed from drying) and pour the mixture over the heap of seed; one or two turnings of the heap will mix the culture satisfactorily and then the seed must be spread out in a shady place to dry. The culture remains active for 8 weeks after issue, but seed should be sown as soon as possible after treatment and must not be exposed to the rays of the sun; therefore, if it is broadcast on the field it should be harrowed in immediately. For the few acres of lucerne usually sown at one time the seed can be treated in a few minutes, at a cost of about 5*s*. per acre. This treatment is a satisfactory result of the application of bacteriology to agriculture, and has done much for farmers growing lucerne in new situations, especially in the north and west of the country.

It is quite possible that inoculation may become common with other legumes, but, at present, knowledge of the various strains of bacteria, and of their prevalence, is insufficient to indicate the possibilities. It may be that wild white clover may be grown more successfully in some places, at considerable altitudes for instance, if inoculation is carried out. Recent work with the bacteria associated with red clover has shown that some strains of these bacteria are merely parasitic, and has suggested that inoculation with picked strains may prove worth while with that crop. As yet, however, it is only lucerne which is inoculated in practice, and the introduction of the treatment has certainly been a valuable step forward in lucerne growing. Where soya beans are grown in this country they must be inoculated, but there seems little chance of this crop ever being satisfactorily acclimatised to Britain.

* The necessary bacteriology was worked out at Rothamsted, and the process has been handed over to Messrs Allen & Hanbury, Ltd., Bethnal Green, London, E.2; the culture can only be obtained from this firm who also supply inoculum for soya beans.

TREATMENT OF POTATO SEED TUBERS

'Sprouting' is a very common treatment of potato setts, the object being to initiate growth before the tubers are planted. The effect is to lengthen the growing period, and in northerly districts this is usually attended by valuable increases in yield; in the southern parts of the country the effect on yield is often negligible, but may be considerable in a wet season, because there is more time for the tubers to develop before the haulms are killed by blight. In any case sprouting ensures that only setts capable of growing are planted and consequently reduces the number of 'misses' in the field. Sprouting is carried out in boxes in which the tubers are laid in a thin layer; the boxes each hold about 30–40 lb. of tubers and the handles at their ends enable them to be piled in tiers, so that there are large air spaces between the layers of potatoes. Seed potatoes are boxed as they become available during late autumn or winter, when they are purchased, or obtained in the riddling of the previous year's crop for sale; the boxing should be carried out if possible 2 months before the date of planting, and it is desirable that there should be frequent inspections of the boxes for the purpose of removing diseased tubers. When potatoes are boxed, and kept in the dark, long, thin, etiolated sprouts are produced, and when in the light, short, thick, green ones; some farmers still prefer to keep tubers in the dark when sprouting, because growth is then quicker, but there is no doubt that sprouting in the light is better. The long thin sprouts produced in the darkness are very easily injured and the bruising inevitable in planting is usually enough to cause them to die right back, so that any gain in time is completely lost. Every year more and more potato growers sprout their setts in the light, and in some districts glasshouses for the purpose are common; these houses are built, where possible, with their long axes running north and south so that equal sunlight may fall on both sides, and they are glass right down to ground level. The risk of freezing is a small one in a glasshouse, but in extremely severe weather an oil lamp is often put in the house. Sprouted setts are planted direct from the boxes because if they are shot into sacks or carts the sprouts will be broken off.

TABLE V. SUMMARY OF SEED TREATMENTS

Crop	Purpose of treatment	Treatment	By whom treated	Frequency of employment in practice	Remarks
Wheat	Control bunt	Organic mercury compounds	Farmer	Almost universal	—
	Control loose smut	Hot water	Farmer	Hardly ever	—
	Protect from birds	Tar, lime, chalk, etc. proprietary compounds	Farmer	Rare	Cannot treat against bunt after treating with tar
	Protect from wireworm	Gamma-BHC	Farmer	Fairly common	Dual-purpose dressings used
Barley	Control leaf stripe; Control covered smut	Organic mercury compounds	Farmer	Common	Should be universal
	Control loose smut	Hot water	Farmer	Hardly ever	—
	Protect from birds	Tar, proprietary compounds	Farmer	Rare	—
	Protect from wireworm	Gamma-BHC	Farmer	Fairly common	Dual-purpose dressings
	Assist germination	Kiln drying	Maltsters	Not in Britain	Very common in Eire
Oats	Control leaf stripe; Control covered smut; Control loose smut	Organic mercury compounds	Farmer	Common	Should be universal
	Protect from birds	Tar, proprietary compounds	Farmer	Rare	—
	Protect from wireworm	Gamma-BHC	Farmer	Fairly common	Dual-purpose dressings
Rye	Control ergot	Brine	Farmer	Rare	—
	Control covered smut	Organic mercury compounds	Farmer	Fairly common	—

	and avoid bunched seedlings		the service		
		Rubbing and grading	Seed firms	Increasing	—
	Protect from wireworm	Gamma-BHC	Seed firms	Common	—
	Control black leg	Organic mercury compounds	Seed firms	Common	Effective but disease may be caused by soil organisms
Peas	Protects from moulds in soil	Organic mercury compounds	Farmer	Rare	Advisable for early sown peas
	Assist germination	Water soaking	Farmer	Very rare	Not uncommon among smallholders
Mangolds	Assist germination	Sulphuric acid	Firms advertising the service	Rare	Probably beneficial in dry season
		Water soaking	Farmer	Fairly common	Effect very variable, depending on soil moisture
Turnips, swedes, kale	Protect from turnip-fly	Gamma-BHC	Farmer	Becoming general	—
Trefoil	Assist germination	Milling	Seed merchants	Very common	Farmers often use own cosh; true seed better in dry spring
Sainfoin	Assist germination	Milling	Seed merchants	Common	Milled seed probably better in dry spring
Clovers	Assist germination	Scratching with wire brushes	Seed merchants	Fairly common	To puncture envelopes of hard seeds
Lucerne	Add symbiotic bacteria	Inoculation	Farmer	Common	Advisable if lucerne not grown on field previously
Potatoes	Start growth	Sprouting	Farmer	Very common	Better done in the light
	Economise seed	Cutting	Farmer	Rare	Only on small farms, except in years when seed tubers are very expensive

Note. Where the Table shows 'farmer' as the one who does the treatment the farmer often, in fact, sends his own seed to a local merchant for treatment or buys seed ready treated.

In recent years there have been interesting developments in Holland in the matter of potato sprouting and Dutch methods are being adopted by a few farmers in this country. The purpose is to produce the right type of sprout, at the date required, by control of temperature and light; humidity also has an effect but precise control of that factor is difficult. Potatoes do not sprout at temperatures below 40° F. and so for the early part of the winter the object is to keep them cool, which is achieved by fans drawing the cool night air into the store. Sprouting is done in a closed building, the light being provided by hanging fluorescent tubes. The potatoes are spread thinly (not more than 30 lb. in a box) in chitting boxes which are piled in tiers leaving alleyways along which the fluorescent tubes can be moved. The lights are not used while the temperature is kept down but some six weeks before it is expected to plant the tubers the temperature of the store is allowed to rise and the lights are used for some 12 hours a day. There is much still to be learnt before the most effective and most economic methods can be defined, but it is already clear that methods will have to be adapted to the varieties of potatoes, which differ markedly in the readiness with which they produce sprouts. If a barn is available the cost of equipment (rather over £10 per ton of storage capacity) is only about half that of building a glass chitting house; the cost of electricity is of the order of 10*s*. a ton of seed a year. In addition the annual charge (interest and depreciation) for chitting boxes is from £2 to £3 a ton of seed.

Any piece, containing an eye, of a potato tuber is capable of producing a plant, and therefore it is possible to economise on seed by cutting each sett into two or more pieces; except in the case of very expensive tubers of a new variety, only one cut is made down the sett in the long direction. In addition to economy of seed it may be urged for the treatment that by reducing the number of eyes in each piece planted it leads to a crop producing a high proportion of saleable tubers, with no diminution of total yield. Smallholders with little to do during the months of January and February commonly spend much time at that season cutting potato setts, but larger growers rarely adopt the practice; they feel that their labour force can be better employed on other work. They also urge that there are more 'misses' in a field planted with cut setts, because a proportion of the eyes in a tuber may be 'blind', and consequently a piece of the tuber may be incapable of producing a plant; this is particularly liable to happen in the variety Majestic. When cutting is performed some care should be taken to protect the newly exposed surfaces. The cut faces become covered with a corky layer, which effectively shields them from infection as long as the layer is formed slowly; if it forms rapidly it cracks and is never a complete cover. The worst procedure is to expose the newly cut surfaces to the rays of the sun; the correct method is to pile the cut setts into a heap which is covered with damp sacks, so that the corky layers may form slowly and remain continuous over the surfaces. It is common to dust lime over the cut faces, but it has not been found in experiments that this has any effect, either beneficial or harmful.

SUMMARY

The seed treatments described in this chapter are very diverse in character, and directed to very different ends. It has been thought well, therefore, to collect them in summary form, in which they are presented in Table v. This table adds nothing new to what has already been said, except that an attempt has been made to appraise the frequency with which each treatment is used in practice. This can only be done extremely roughly, but some indication may be useful to the reader. The whole subject of seed treatment is of absorbing interest and offers rich returns; it is a subject which is developing rapidly and which may be expected to make further big strides in the future.

In this chapter no mention has been made of 'vernalisation', a seed treatment of which much has been heard; it has not been included in Table v because it has played no part in British farming. The principle of the treatment is that there are certain developmental phases through which the young plant has to pass, and that time may be saved in growing the crop by taking the seed through these phases before it is sown. Thus winter wheat, if sown in spring, may produce no ears at all, but if the seed be soaked in water and held at a low temperature for several weeks, it may behave as a spring variety. In some districts of Russia, where the treatment has been developed, it has been widely used as a means of overcoming climatic conditions unfavourable to wheat, and also with other crops. In this country there appears to be little hope that vernalisation will prove of benefit to the farmer, though uses may be found for it in market gardening.

CHAPTER VII

SOWING THE SEED

There has been great debate over the effect of plant population on yield. Much experimental work has been conducted, particularly with wheat and sugar beet, and the problem has proved to be very involved; considerable progress has been made, however, especially in the case of wheat. Clearly the problem differs, as it affects sowing, between singled crops like sugar beet and unsingled crops like wheat; in the latter case the final plant population on the field is closely related to the amount of seed sown and to the methods of sowing, the relationship being modified, of course, by the conditions of soil and season. It is proposed to discuss the unsingled crops first, and though wheat will be the crop chiefly mentioned the problem is probably very similar with the other cereals.

There is an old doggerel which runs:

> Sow four seeds in a row,
> One for the pheasant, one for the crow,
> One to rot and one to grow.

This is unduly pessimistic, as a plant establishment of only 25 per cent of the seed, is happily, infrequent. Birds sometimes take a considerable proportion of the seed sown, but it must not be thought a general rule that they will get half of it; with late-sown wheat their depredations may occasionally account for as much as half, though in that case the chief offender is neither the pheasant nor the crow, but the humble starling. It does not always follow that a plant is lost when the seed is taken, because starlings and larks levy the greatest toll when the young plants are just coming through the ground; undoubtedly many plants die when robbed of the seed at this stage, but it is often surprising to observe how many survive, although their root systems appear very rudimentary at the time. If proper care is taken in selecting and treating seed the proportion which rots should certainly be much less than one in four, though subsequent attacks of pests, such as wireworm, may seriously reduce the ultimate survival. Counts of plants on the field show great variation in the proportion of the seed contributing plants to the crop, but a plant establishment of as low as 25 per cent is rarely suffered, and such crops are often partial or complete failures; figures of 50 per cent would be more normal, whilst establishments up to 70 per cent of the seed sown are not uncommon. It is quite important to appreciate that a normal figure would be 50 per cent rather than 25 per cent or even 10 per cent, as is sometimes stated. If the last were the true normal figure there would be little point in arguing about seed rates, because the only logical thing to do would be to crowd on enough seed in a blind effort to provide the field with a reasonable covering of plants; but if a 50 per cent germination may be expected the case is very different, for then the farmer has considerable control on the plant population, in the number of seeds he sows and in the regularity with which he sows them.

Exhaustive studies have revealed that, speaking still of wheat, there is a marked propinquity effect, that is, that the yield from a very small area is largely affected by the competition its plants have to meet from surrounding areas. If the plants are counted, and the yield determined, for a number of small samples (such as 1 ft. length of a coulter row) it is found that yield increases with increasing number of plants up to an optimum population of about 16 or 17 plants per foot, and then decreases; but this is probably an overestimate of the optimum, because the high yields of thick foot lengths are produced largely at their neighbours' expense. It follows that the optimum plant population can only be determined by sowing a considerable area accurately at a constant seed rate, so that the propinquity effect may be roughly the same all over it. The figures in Table VI were obtained from such an experiment, where the seeds were sown by hand as evenly as possible on every foot-row of the plots. The field was in an exceedingly rough state of tilth at the time of sowing, and lay very wet throughout the following winter, but even then, it will be observed, the effective germination was nearly 50 per cent. Lines 5, 6 and 7 of the table show how compensatory effects come into play, and tend to even out the differences between the rates of sowing. On the thicker plots each

TABLE VI. DEVELOPMENT AND YIELD OF WHEAT SOWN WITH VARYING NUMBERS OF SEEDS PER FOOT OF COULTER ROW

(1) No. of seed sown per foot	6	10	14	18	22	26
(2) Equivalent seed rate (bushels/acre)	0·55	0·91	1·27	1·64	2·00	2·36
(3) Effective percentage germination	44·6	46·9	50·2	50·5	48·8	46·8
(4) No. of plants per foot	2·7	4·7	7·0	9·1	10·7	12·2
(5) No. of tillers in April per plant	3·8	3·0	2·7	2·6	2·1	2·0
(6) Proportion of April tillers bearing ears	0·83	0·73	0·68	0·62	0·64	0·64
(7) Wt. grain per ear (g.)	1·47	1·34	1·20	1·10	1·03	0·93
(8) Yield of grain per foot (g.)	13·07	13·88	15·10	15·35	14·50	14·05
(9) Equivalent yield per acre (bushels)	37·5	39·8	43·3	44·0	41·6	40·3

plant produced less tillers, each tiller had less chance of producing an ear, and each ear contained less grain. And so it was that the highest yield was not obtained from the most thickly seeded plots, but from those sown at the equivalent of a little over 1½ bushels per acre, which, in the rough sowing conditions, meant only nine plants on a foot of coulter row. No general lesson can, of course, be learnt from any one experiment, but for heavy clay soil the figures shown in Table VI may be regarded as typical; the optimum population may vary somewhat between soil types, but as yet no definite statement can be made on this, except that the present indications are that more plants are required on light chalky soil than on heavy clay. In all cases there is an optimum plant population, and the question arises whether it is worth while striving, by working the land to a fine tilth and by finding a drill that will sow really evenly, to ensure that every foot-row on the field carries the optimum number of plants. This would undoubtedly be a desideratum were it not for the propinquity effect referred to above; it appears that this effect is large enough to obliterate the harm done by small irregularities in sowing, but obviously its power to do so is limited. If a drill coulter is momentarily blocked, or a clod is encountered and hence plants are missing for a few inches of drill row, yield will probably be unaffected because the surrounding plants, in and across the row, will benefit from lack of competition sufficiently to compensate for the gap. But if areas of the field of the size of a square yard or more are devoid of plants, neighbours cannot compensate for the gap; what is probably more serious is that on the bare ground weeds will flourish. It is impossible to define accurately how uniform a 'plant' should be; the farmer must obviously prepare a reasonable tilth and must discard a drill before it has reached the last stage of decrepitude, but it seems doubtful if any appreciable increase in yield can be obtained by regularity above the present normal. It is fairly common to see, at regular intervals across a field, rows of plants which are thin or missing. This gives the field an unsightly appearance and is usually due to a defective coulter of the drill (e.g. one worn sharp and cutting in too deeply); probably the loss in yield in such a case is quite small, but if two adjacent coulters were defective the loss would be serious.

Cereals tiller at a certain stage of growth, that is to say, a form of branching occurs at about ground level, so that a number of stems, often as many as twenty, are produced by each plant; each tiller develops a root system, so that each plant becomes a little colony of tillers, each of which is practically independent of the others. Autumn-sown wheat should be up with rows showing green before Christmas, and tiller production may commence any time after that, being continued until early May. Farmers consider tillering to be very important, and it certainly helps in that it leads to more than one ear per plant. But it may be a snare and a delusion, for only a small proportion of tillers survive to bring forth ears for the harvest; thus though the number of tillers per plant in the first week of May may be as high as twenty, the number of ears per plant at harvest is not often more than two, since from mid-May onwards the majority of the tillers die back, only existing as debris at harvest. It is reasonable to hope that a gappy 'plant' may tiller more than a full one, and though this is generally true, the farmer may be lulled into a false sense of satisfaction by this very fact. Careful studies have shown that the increased tillering caused by thinness of 'plant' does not manifest itself until the end of March; it has also been shown that it is the main stem and one or two of the earliest tillers which provide the ears at harvest. Therefore when a field looks gappy during winter, but tillers abundantly in April, satisfaction in seeing the field become covered is ill founded, because very few of the extra tillers will contribute to the yield of grain.

From the above general survey of the question it may be concluded that plant establishment is usually a higher proportion of the number of seeds sown than is generally believed, and consequently that seed rate is important. The practical view that there is an optimum seed rate is found to be borne out in experiments, but these usually show that the optimum is lower than the rate which would be regarded as normal in practice. A reasonably even plant is desirable, but special efforts to ensure meticulously even seed distribution are unlikely to bring a reward. On the other hand, an abnormally gappy 'plant' will produce a relatively low yield because, although the gappiness may be hidden by abundant tillering, the tillers will be formed late, and will have little chance of bearing ears. With other cereals the position is apparently very much the same, and with spring-sown cereals it is unlikely that any new principles are involved, the only difference being that the developmental changes are telescoped into a shorter period. It is possible indeed that similar views are justified in the case of beans, with which the optimum number of plants per unit of row appears to be about one-third that in the case of wheat; though beans do not tiller, they branch above ground level.

The seed rates commonly found in practice are given in Table VII, which shows the limits within which most cases in this country would lie. With wheat almost always, and with other crops generally, experiments suggest optimum seed rates lower than those used in practice, but it is risky to advise a general reduction; seed may be of inferior quality, and conditions are frequently unfavourable, so that the farmer's pessimism in providing more than should be necessary is quite commonly justifiable. The points which influence a farmer in modifying his seed rates are as follows:

(*a*) *Other factors may limit the yield.* The best examples of this must be sought in those countries where rainfall normally limits the yield to a very low figure; thus in parts of Australia the normal seed rate for wheat is only 1 bushel per acre. In this country there is no general variation according to rainfall, except that on the very light soils of the Suffolk breckland seed rates are very low, and 1½ bushels for wheat would be considered adequate. This is, of course, very poor wheat land, and crop failures are frequent.

(*b*) *Richness of the soil.* On very fertile soils it is customary to sow rather less than normal, on the principle that the chief danger is that the crop may be too thick and rank, and thick crops are more liable to lodge than thin ones.

(*c*) *Condition of tilth.* Seed rate is usually raised somewhat if sowing is done on a very rough tilth, because, owing to variation in the depth of sowing, an abnormal proportion of seeds will not grow.

(*d*) *Method of sowing.* Where the seed is broadcast, more must be sown than where it is drilled, again because of the uneven covering it will receive, and consequent low plant establishment.

(*e*) *Time of sowing.* For early sowing less seed is recommended, because there will be more time for tillering and more chance that tillers may be produced early enough to bear ears. In the extreme comparison of autumn and spring sowing the point is undoubtedly important, and an increase of a bushel per acre for the latter is probably desirable with all cereals and beans.

(*f*) *Size of seed.* The object is to sow a certain number, not weight, of seeds per acre, and consequently varieties with small grain require a lower seed rate than others, because each bushel will contain more seeds.

(*g*) *Birds.* If sowing is late in autumn or early in spring, or if past experience has been unfortunate, birds must be expected to filch a proportion of the seeds sown, and consequently it is usual to increase seed rate. The principle is a reasonable one and is usually justifiable, but in really bad cases sowing more seed is merely feeding the birds more generously.

It will be seen from Table VII that for nearly all corn crops the normal seed rate is 2 or 3 bushels per acre. The exception is oats, for which the seed rate is about double that for the others. Winter beans have the same rate as the cereals because there are two differences to be considered which nullify each other in the final result; the seed is much larger than that of cereals, but the crop is drilled in rows much wider apart. The difficulty with beans is to get the right population—many crops are so thin that a full yield cannot be expected, whilst many other crops look most promising during growth, but are too thick to carry many pods, so that threshing is miserably disappointing. The general view of farmers is that 2 bushels are sufficient, but as rooks are inordinately fond of beans another bushel is usually sown. Recent experiments have given results in favour of higher seed rates and as much as 4 bushels an acre has been recommended. With peas the seed is smaller than with beans, but the rows are drilled closer together so that the seed rate remains unchanged. Rye and tares are often sown with the object of feeding them in the green state; in that case thick crops are required and the seed rate should be increased to 4 bushels. A few other crops which play a small part in British agriculture have been included in the table for reference.

Turning to root crops, most of which are singled, the considerations are different, because the final population depends on the distance between the rows and the distance to which the plants are singled in the row. With them, too, however, there is an optimum seed rate. On the one hand enough must be sown so that the singlers will find a continuous row of plants, whilst on the other hand, if the seed is too thick the plants will come up spindly and wound round each other, so that it is impossible for the singlers to remove some without disturbing those they leave. Sugar beet is the root crop most studied in this respect, and the general view is that absolute regularity of plant must be diligently sought; whilst experiments have shown that with sugar beet the 'propinquity effect' is strong, and that single gaps may be fully compensated by the extra growth of neighbouring beet, it is quite certain that more loss is suffered in practice by fields being too gappy than by the roots being too thick on the ground. In the case of Brassicae the whole position has been changed by the introduction of seed dressings which effectively protect the young plant from the fly. An ample seeding was favoured in the hope, usually vain, that enough plants would survive the attack of the fly. When the seed dressings were first used many farmers found that they were getting the plants too thick in the row;

TABLE VII. SOWING DETAILS OF VARIOUS CROPS

Crop	Normal seed rate (per acre)	Range of rates in practice	Normal distance between rows (in.)	Range of row distances in practice (in.)	Optimum depth of sowing (in.)	Normal singling distance (in.)	Range of singling distances in practice (in.)	Remarks
Wheat	2½ bush.	1½–4 bush.	7	6½–10	2	—	—	—
Barley	2½ bush.	2 –4 bush.	7	6½– 8	2	—	—	—
Oats	4½ bush.	3 –7 bush.	7	6½–10	2	—	—	Size of seed very variable between varieties
Rye	3 bush.	2 –4 bush.	7	6½–10	2	—	—	Higher rates of seeding if for green feed
Beans	2½ bush.	2 –4 bush.	20	18 –27	2½	—	—	—
Peas	2½ bush.	2 –3 bush.	10	7 –15	2	—	—	—
Tares	2½ bush.	2 –4 bush.	7	6½–10	2	—	—	Higher rates of seeding if for green feed
Lupins	1½ bush.	1 –2 bush.	20	18 –24	1½	—	—	—
Linseed	1¼ bush.	1 –1½ bush.	7	6½– 8	1	—	—	For flax closer drilling is favoured so that there may be less branching
Buckwheat	1½ bush.	1 –2 bush.	7	6½– 8	1½	—	—	—
Maize	2 bush.	1 –2½ bush.	24	20 –36	3	—	—	—
White or yellow turnips	3 lb.	1½–6 lb.	20	18 –27	½–1	10	7–14	White turnips sometimes broadcast in September, and not thinned
Swedes	3 lb.	1½–6 lb.	20	18 –27	½–1	12	8–15	—
Kohlrabi	3 lb.	1½–6 lb.	20	18–27	½–1	12	8–15	—
Kales	3 lb.	1½–6 lb.	20	18 –27	½–1	10	8–14	Often not singled

Rape	3 lb.	1½–6 lb.	20	15–24	½–1	—	—	Seed rate up to 10 lb. if broadcast. Not singled
Mustard	10 lb.	7–20 lb.	12	10–20	½–1	—	—	Often broadcast. Not singled
Mangolds	8 lb.	6–10 lb.	22	18–30	½–1	14	10–17	—
Sugar beet	15 lb.	12–25 lb.	20	15–24	½–1	10	6–12	Seed rate halved for rubbed seed
Cabbage	1 lb.	½–1 lb.	24	20–36	½–1	18	15–30	Seed rates are amounts in seedbed to give plants per acre; 3 lb. if drilled on field
Potatoes	17 cwt.	6–25 cwt.	27	24–30	3	15	12–21	Depth is given as distance below mean level of field
Red clover	14 lb.	8–20 lb.	7	3½–8	½	—	—	Often broadcast
Trefoil	14 lb.	8–20 lb.	7	3½–8	½	—	—	Often broadcast Double seed rate for cosh
Trefolium	20 lb.	10–30 lb.	7	3½–8	½	—	—	Usually broadcast on stubbles
Lucerne	20 lb.	15–25 lb.	7	3½–8	½	—	—	Sometimes broadcast
Sainfoin	56 lb.	50–60 lb.	7	3½–8	½	—	—	Seed rate doubled if seed unmilled
Kidney vetch	20 lb.	16–22 lb.	10	8–14	½	—	—	—
Rye grass	24 lb.	20–35 lb.	7	3½–7	½	—	—	—
Pasture mixtures	30 lb.	10–35 lb.	7	3½–7	½	—	—	Broadcasting common

some drills will not sow less than 4 or 5 lb. an acre and to get lower seed rates the seed is sometimes mixed with bran. It is important to note that seed rates for root crops are matters of pounds and not bushels, as in the case of cereal crops. When actually engaged in farming one soon gets to know the usual seed rates, but the tyro may find it useful to remember that the normal rate is approximately 4 lb. per acre for Brassicae, double that for mangolds, and is again doubled for sugar beet. It is scarcely worth while burdening the memory with minor departures from the above rule, because sowing conditions (particularly in regard to moisture) may vary the percentage establishment considerably, and, in any case, experimental evidence on the effect of slight departures from the normal is lacking. With turnips, swedes, kohlrabi and kale the tendency is to sow rather more in the north than in the south of the country. Rape is often broadcast, when the rate is raised appreciably, sometimes up to 10 lb. Mustard has a high seed rate because the plants have not a great lateral spread and it is desired to cover the ground thickly; 10 lb. would be normal for drilling and 20 lb. for broadcasting but these rates can be reduced considerably if the seed is dressed against the fly. The increased rate for mangolds is due to the fact that it is a fusion of fruits which is sown, and this is much larger than the true seeds of Brassicae; sugar-beet fruits are very like those of mangold, but the former produce smaller roots and therefore more of them are required per acre, and compensation for thinness of 'plant' is more limited. Much lower rates are used where rubbed seed is sown and where a young drill is employed. Cabbages are sometimes drilled direct on the field, to be singled later, and in that case the seed rate is about the same as for turnips, etc.; more often, however, cabbages are sown in a seedbed, the young plants being transplanted later, and in that case 1 lb. sown in the seedbed should provide sufficient plants for an acre. This latter allowance, usually proves a generous one, and where expensive seed (as in the case of new selections) is sown, the allowance is commonly cut down to ½ lb., or even less, which would be regarded as a more normal amount for brussels sprouts, cauliflower, etc., in market gardening. There is very great variation in the weight (in this case usually expressed in cwt.) of potato tubers required to plant an acre. Row widths differ slightly, and distance between tubers in the row appreciably, and both these variations affect the weight sown; but the main cause of the wide spread of rates shown in Table VII is variation in size of sett.

Temporary pastures are often sown with mixtures, but the seed rates given in Table VII are for those species which are often sown alone. In most of these cases the seeds are very small, but the plants are also small and a very thick cover of the ground is required. For most of them the usual rate is from 10 to 20 lb. per acre, but the student is warned of the case of sainfoin, which is a well-known examination catch; sainfoin seed is much larger than clover seed and its rate of sowing is about four times as great, even when the milled or true seed is used.

DISTANCE BETWEEN AND WITHIN ROWS

The first thing to appreciate in connection with row distance is that fine adjustments in the matter are not usually possible. Drills are constructed to run 'wheel to wheel' (in some cases they overlap and run 'wheel to coulter mark', but the limitations are similar in both cases), that is, one wheel runs on the wheel mark made on the previous passage of the field; this clearly necessitates that the distance of either wheel from the coulter next to it should be half the distance between any two coulters. Row distance is varied by increasing or decreasing the number of coulters in the drill. A common width of horse drill is 7 ft. 6 in. and it is sent out by the makers with a complement of 14 coulters. If all the coulters are used the row distance will be 90/14 or, approximately, 6½ in.; if 13 are used the row distance will be 7 in., if 12 are used 7½ in. and so on. Thus it is impossible to get the rows nearer together than 6½ in. and in widening them it is only possible to proceed by discrete steps of approximately ½ in. When the drill is used for root crops the steps become much larger; 6 coulters give a row distance of 15 in., 5 give 18 in. and 4 give 22½ in., and a drill 7 ft. 6 in. wide cannot give the intermediate row distances. Theoretically it is possible to arrange for the intermediate ones, by a setting which allows for a given overlap (or 'wrapping') of the breadths on successive passages;

but it is very important for after-cultivation that the rows should be constantly and accurately spaced, and in practice that can only be achieved by running 'wheel to wheel'.

With cereals the usual practice is to have all, or nearly all, the coulters in the drill, and thus to put the rows as close as possible. In the case of oats, trials conducted by the West of Scotland College of Agriculture have shown that any increase in row distance beyond 6 in. is accompanied by a small, but definite, decrease in yield of grain and straw. With barley, a series of experiments extending over 4 years was carried out at the Norfolk Agricultural Station, comparing row distances of $3\frac{1}{2}$ and 7 in., the former being obtained by taking the drill over the plots twice; the narrower rows consistently yielded more than the wider ones, the average increase in grain yield being 3 bushels per acre. With wheat, however, similar results were not obtained, no advantage being found for row distances less than 7 in.; this conforms to the farming belief that wheat should be fairly thick in the row. Except for barley there is no evidence then, that anything is to be gained by decreasing the row distance below the normal 6 or 7 in.; at present, it will be realised, there are no drills available which can sow corn nearer than this. It was sometimes advised that rows should be wider than normal so that horse hoeing could be carried out, and 10 in. was often given as a minimum for this operation; horse hoeing was possible, however, with row distances of 7 in., if care was taken in the setting of the hoe and in working within drill breadths, so that all the rows hoed at one time were strictly parallel. Horse hoeing of corn is one of the refinements of farming which do not accord with the prevailing economic conditions; the tractor cannot well replace the horse for this particular purpose and hence the practice has declined markedly in the last three decades.

Beans are drilled much wider than cereals because the stems bear flowers along most of their length, and pods will not be set unless light reaches the flowers. It is sometimes said that beans are drilled in rows wide apart so that they may be thoroughly horse or tractor hoed, but this is not the correct line of argument; they are drilled wide to admit light to the flowers, and, having been drilled wide, full advantage should be taken of the good opportunity thereby offered for hoeing and continuing the operation relatively late in growth. Wide drilling is not by any means universal for beans at the present time and cultural methods are being adopted which have as their main object the elimination of all hoeing (see p. 234). Peas are often drilled at distances intermediate between those of cereals and beans. They have weak stems and consequently 'flop' on the surface of the ground soon after they are a foot high; since they then fail to smother weeds it is important that they should be thoroughly hoed whilst still erect, and this is the reason for the common practice of drilling them wider than cereals. As with beans, so with peas, hoeing is much less common than formerly and frequently they are drilled with all the coulters in the drill to give rows as close together as possible.

Of singled crops sugar beet is the one whose spacing has been subjected to most enquiry. It is usual to speak of plant populations of beet in terms of numbers of roots per acre, and 30,000 appears to be about the optimum number; rows 18 in. apart, singled to 12 in. between the beet in the row, give a possible population of 29,040. Many experiments have been made on the question of the optimum number of beet per acre, and it must be confessed that the results have been far from consistent; a general survey of them leaves no clear impression of definite differences in yield associated with variations in plant population over as wide a range as 20,000–40,000 per acre. It is often argued that discrepancies in results should be attributed to differences in fertility level. There is one school of thought which holds strongly that more beet are required per acre on poor land than on rich land, but this must be regarded as not proven; in fact, such experimental evidence as there is bearing on this point indicates that the optimum population is not different on different soils. Possibly the reservation should be made that on very light land more beet than normal should be grown, so that the ground is covered earlier and the danger of the soil being blown away by the wind lessened. On the continent of Europe, where sugar beet has been grown for more than 100 years, the practice is to have the rows close together (about 14 or 15 in.) and to single fairly widely in the row, so that the beet are almost equally spaced in the row and across the row. But labour is much cheaper in the continental sugar-beet districts than in this country, and so when the crop was introduced here the natural tendency was to have the rows wide apart, so that more row-crop work and less hand hoeing could be carried out. There has been an attempt to compensate for the greater row distance by growing sugar beet close together in the row, singling distances of as little as 6 in. being used; this compensation, however, does not occur, and there is plenty of evidence that nothing is to be gained by singling closer than 10 in., whilst 12 in. is probably not too wide. Practice is by no means standardised in regard to distance between the rows, but opinion is hardening against widths greater than 21 in., and this is supported by the majority of (but not all) experiments. So many considerations have to be taken into account in deciding this apparently simple issue. Some farm horses have very big feet, whilst others pick them up as though they wanted to show off their shoes to a bystander, and this splaying action means that they require a wide path; consequently if horse hoeing is to be carried out as late as possible in growth, a row distance much below 18 in. must be avoided. It must also be appreciated that with all root crops, if the field is unduly foul, it is wise to put the roots a little wider than usual, so that row-crop work may be continued later in growth. Row-crop tractors have narrow wheels and easily adjustable wheel-spans but in practice they are subject to much the same limitations as horse hoes. Narrow rows mean more beet to the acre and, since yield is not greatly affected, this means that the individual beet are smaller. Much has been made of the point that smaller beet have a higher sugar content than larger ones, and there is no doubt that this generalisation is true; but the difference in sugar content over the normal range of size of root is small. When it comes to harvesting it is very much quicker, and hence cheaper, to handle relatively few large beet than relatively many small ones. Naturally these points appeal to farmers with varying force and hence the wide differences found in inter- and intra-row distances between even the successful beet growers; the fact that considerable latitude is possible over plant population without marked effect on yield has already been noted. It is not possible to make any definite statement to which no growers will take violent exception, but the following appear to the writer to be true:

(1) The best row distance is about 18 in., and this should certainly not be increased beyond 21 in.

(2) There is little to be gained by singling closer than 10 in., and 12 in. is probably near enough.

(3) The above is applicable to all soil types, with the possible exception of the lightest, blowing sands.

The mangold is larger than the sugar beet, and consequently it is normally drilled and singled at a slightly greater distance; fodder beet have smaller roots and they can probably be grown most successfully at about the same spacing distances as sugar beet. In point of fact many crops of mangolds and fodder beet are small in area and often the drill has just been used for sugar beet, and the setting is left unchanged. Turnips, swedes and kohlrabi are spaced very much as recommended for sugar beet, though row distances are inclined to be greater; where they are grown on ridges the latter cannot be placed nearer than 24 in. It is curious to note that farming opinion favours closer rows on rich land to prevent the roots from becoming coarse; the reader will appreciate that this principle is precisely the reverse of that so often advanced in connection with sugar beet. The kales are drilled with row distances similar to the above, but a difference arises in regard to singling. Where marrow stem kale is wanted for green soiling it is usually singled to 8 or 10 in., because it is easier to cut and cart a few large stalks than many small ones; where the kale is intended for sheepfolding, however, it is very often left unthinned, which greatly cheapens the cost of growing it. Experiments comparing thinned and unthinned kale contradict each other so diametrically that it can only be concluded that there is little difference in yield. Some farmers compromise on the point and have their kale chopped out so that little bunches of three or four plants are left at distances of about 7 or 8 in. along the rows. The distances to which cabbages are spaced, at transplanting or at singling, are determined by their type; large Drumhead varieties are often planted 'on the square' as far apart as 1 yd., but newer strains of small cabbage for human consumption are spaced little more than mangolds. Maincrop potato ridges are usually 27 or 28 in. apart, this distance being very convenient for the ridging plough and fitting the wheels of farm carts or trailers; early potatoes, or potatoes grown to provide seed tubers, may be planted in ridges as close as 24 in. On foul land ridges may be farther apart so that cleaning may be continued later in growth; ridges are also sometimes put wider than normal when a large-topped variety like Kerr's Pink is being planted. The distance from plant to plant in the row might probably be increased on most farms with advantage. It has been shown that there is little difference in total yield between spacings of 12 or 20 in. in the row, but that at the longer distances the crop yields a larger proportion of ware, or saleable, tubers.

If clovers and grasses are drilled they are sown in rows as close together as possible. The importance of close rows is sufficient to warrant a special drill for sowing small seeds; some old 'seeds' drills, with coulters $3\frac{1}{2}$ in. apart are still giving good service and new ones with coulters at 4 in. are being produced. Where no such drill is available it is not very uncommon with leys to drill the field twice, at right angles. As will be seen later, the seed of clovers and grasses is very often broadcast, and a point in favour of that method is that it gives a better lateral distribution of the seed than can ever be obtained by drilling.

DEPTH OF SOWING

The considerations which arise in regard to depth of sowing are the following:

(*a*) *The seeds must be covered.* In the case of corn crops this is very necessary so that they may be, to some extent, protected from the birds. Smaller seeds will often germinate successfully on the surface of the ground, but this depends entirely on an adequate intermittent rainfall, until they have established plants with fairly well-developed roots.

(*b*) *The seed must be in damp soil.* The danger that seeds may germinate and then find insufficient moisture to keep them alive is a very real one. In spring or summer sowing, when fine weather may be expected subsequently, this point is important, and may reasonably decide a farmer to sow a little deeper than usual.

(*c*) *On a rough tilth seeds should be sown deeper than on a fine one, because otherwise a large proportion will not be covered at all.* This provides the reason for preparing a fine tilth when sowing small seeds, because the smaller the seed the less its food reserve, and hence the nearer it must be sown to the surface.

(*d*) *The seed must not be too deep.* When sowing is unduly deep many plants fail to get through at all, whilst those which struggle to the surface emerge in an etiolated and weak condition. Such crops rarely yield well, and, in the case of cereals, are very weak strawed and liable to lodge.

The usual depths for the various crops are given in Table VII, but a word on the difficulty of deciding on what the depth actually is must be said. It is usually easy to find the bottom of the mark made by the drill coulter, and even to find the seed after it has been sown, but then it is by no means easy to measure the depth of sowing; the surface in the immediate vicinity is bound to carry a number of clods, and though these may be small they will be sensible in size compared to the distance it is required to measure. Two people measuring depth of sowing in the same case can easily disagree by as much as an inch, so that the figures for depth in the table must be taken as only approximate. With clover and grasses all that is wanted is a light powdering of soil over the seed, but this can scarcely be measured as less than half an inch.

METHODS OF SOWING

There is a variety of methods available for sowing seed, and these must now be discussed.

(I) BROADCASTING

This method consists of casting the seed as evenly as possible over the surface of the ground, covering being obtained usually by harrowing, and occasionally by just pressing the seed into the ground with the roll, or even by treading with sheep. Broadcasting may be by machine or by hand, the latter being probably the more common, though it is declining as men with experience of the art are becoming rarer. In hand work the seed is carried in a container (usually of metal and called a 'seed-lip' or 'seed cord') slung across the front, or slightly to one side, of the stomach from a strap going over one shoulder. The width covered is almost invariably $\frac{1}{2}$ rod (that is, $8\frac{1}{4}$ ft.), and guiding posts are placed in the headland towards which the man walks (intermediary posts to give the line being used on long fields); at each turn the post at that headland is moved across double the working width to give a mark for the next but one journey across the field. Seed is cast 'with the step', some men using only one hand, but most men using both; a better distribution is obtained by throwing the seed upwards and forwards, though it must not be thrown very high in windy weather. An experienced man walks with a good swing and is remarkably accurate in the amount of seed he sows per acre; distribution, however, is not usually very even, the tendency being for more seed to be thrown straight in front of the path of walking than on either side.

There are three machines which can be used for broadcasting seed. The 'fiddle' is carried by a man and contains a hopper from which a thin stream of seed falls on to a horizontal disk which throws it out, since the disk is rotated by drawing a bar backwards and forwards over a cogged spindle; the action resembles somewhat that of a violinist with his bow. The seed barrow is very commonly used for broadcasting grass and clover seed, and most farmers possess one. It consists of a box—usually 9 ft., though sometimes 12 ft., long—containing the seed and carried on a framework like that of an ordinary barrow; a spindle, carrying brushes at intervals along its length, runs through the box and is rotated by gearing connecting with the wheel on which the barrow runs; the brushes sweep the very small seeds to holes in the lower rear face of the box, through which they fall to the ground. Seed rate is varied by moving tin plates, and thus varying the number of holes left uncovered; with a little experience the amount of seed can be determined accurately, and the distribution over the width of working is very even. Finally, it is possible to use an ordinary drill for broadcasting, taking it over the field with the delivery mechanism in action, but with the coulters raised clear of the ground. There is no doubt that the last method and the seeds barrow give a much more even distribution of seed than can be achieved with the fiddle or in hand sowing unless the latter be done by an expert. Broadcasting has already been referred to in Chapter v as a good means of sowing seed on land lying in the plough furrow, and the method is quite frequently employed on a farm scale with oats. If the season is getting late, seed may be broadcast on unploughed land and covered in with a very shallow furrow, but the resultant crop is often very poor. Broadcasting is sometimes used by smallholders and by larger farms for odd pieces of land, and is, of course, frequently employed for sowing chemical fertilisers. In its favour may be urged the fact that it is cheap and quick and requires (with the exceptions given above) neither machines nor power. It might also be urged that the seed is better distributed laterally than can be achieved by drilling, because in the latter the seed is confined to rows. It is doubtful, however, whether this general cover of the ground is of any importance with cereals, whilst with pulse crops it is better to confine the seed to rows so that light may reach the flowers, and with root crops it is vital in order that hoeing may be performed. With temporary and permanent pasture the point has some weight, though with close, or double, drilling the lateral spread may be sufficient for practical purposes; in the drier districts of the country it is considered more important to put the seed into the ground than to get a good lateral distribution, but over most of the country broadcasting (usually by seed barrow) is the normal method of sowing clovers and grasses. The obvious and serious disadvantage of broadcasting is that the seed is only put on the top of the land, and its covering depends on the haphazard action of harrow tines; much is often left on the surface and is lost to birds, or fails to germinate because of lack of moisture. It has already been seen that lowered percentage plant establishment with the method necessitates a higher seed rate. Further limitations are the inability to hoe and the fact that a skilled man is needed to do the work well by hand, and such men are becoming ever harder to find.

(2) DIBBLING

This method of sowing seed is rarely found on a farm, the garden being its usual sphere of usefulness. During the early years of this century it was still possible to find a few farms on which large acreages of beans were sown by this method, but the rise in the price of labour has made that utterly uneconomic; in its favour might be urged the great economy in seed and the good spacing of the plants. On the farm dibbling is, of course, still common for transplanting cabbages and other crops of the market-gardening type, though efficient transplanting machines are on the market. It is important that the soil should be pressed firmly round the plant roots, and this is effected in transplanting machines by heavy wheels running on each side of the row. In hand work a hole is made, the plant inserted and the dibbler thrust into the ground close to it, to bring the soil firmly against the lower part of the root; an alternative method is to have a gang of men working with an ordinary plough, the plants being laid against the last turned furrow, and 'heeled in' after the next furrow has been turned on to them.

Under the heading of dibbling, planting of potato seed tubers might be considered, though in that case a separate hole is not made for each tuber; the tubers are, of course, dropped singly, by hand, into the furrow opened by the double-mould board plough, covering being effected by splitting the ridges to form new ones over the setts. This was the universal method before the introduction of efficient potato planting machines. These work on a flat seedbed, a small double mould board running in front of each delivery coulter to open a furrow for the sett; there follow mould boards to cover in the setts and leave them in ridges which are usually rather shallow. Some machines incorporate a hopper for chemical fertiliser which is delivered down the same coulters as the setts; it has been seen in Chapter II that a desirable development would be sowing the fertiliser in bands to the side of the seed tubers. Potato planters economise labour and have one major advantage over hand planting in that the furrow is opened and closed at one passage of the machine; this avoids what may prove a serious loss of moisture between drawing out ridges and splitting them, an interval of hours, or even days, where the older methods are used. Potato planters would soon become almost universal if an equally satisfactory machine could be produced for harvesting the crop. Most farmers still rely on hand pickers for this and these are usually casual workers, mostly women. In potato-growing districts it is difficult for a farmer to get a gang of pickers if he has not offered casual employment in the spring of the year for planting the crop. Thus, though potato planting has been successfully mechanised the full exploitation of this awaits equal success in mechanising potato harvesting.

(3) PLOUGHING IN

It has been mentioned that seed may be covered with a light furrow, but then the seed is first distributed broadcast by hand, and that method has already been described. Under the present heading the only crop that has to be considered is beans (very occasionally maize) for which a special little drill is available for fixing between the stilts of a horse plough or on the frame of a tractor plough. The beans are dribbled out of the drill on to the furrow bottom which has just been opened, and are covered by the next furrow slice turned. The method is frequently employed and the following may be claimed as its advantages:

(*a*) The bean drill is very inexpensive and in use it hardly affects the rate of ploughing, since it can be switched on and off without stopping.

(*b*) Cultivations are saved. When the field is ploughed it is sown and there is no question of having to work down a seedbed. In a wet autumn, seedbed preparation on heavy land is difficult, and the land is often dry enough to plough when it is too wet for the use of cultivator and harrows. In such seasons work necessarily gets behind, and it is more than a convenience to be able to push ahead with one crop, whilst waiting for conditions which permit the drilling of wheat.

(*c*) The seed is well covered and moderately safe from birds, rooks being very fond of beans.

(*d*) The seed is on a 'firm bottom'. Beans are reputed to like this, but the writer knows of no satisfactory evidence that they grow better when placed on the top of soil which has been plastered down by the heel of a plough.

The disadvantages of ploughing in beans are:

(*a*) That the land is not tilled.

(*b*) The seed is usually sown too deep. Beans are large and can safely be sown as deep as 3 in., but the depth when they are ploughed in is more often 5 or even 6 in. The result is that the plants which come through are yellow and weakly, whilst often many of them never succeed in reaching the surface at all, especially when the ploughing was done in wet weather to give glazed furrow slices, setting very hard on drying.

(*c*) The rows of plants will not be so straight as when drilled. The young bean plants all tend to come through the ground at the point where one furrow slice leans on its neighbour; consequently although the seed is liable to be spread over a width of up to 10 in., the plants do come up in definite rows, but the plough furrows, and hence the rows, are apt to vacillate somewhat. Straightness is very important because without it good horse or tractor hoeing will be impossible.

Speaking generally, crops that are ploughed in are usually poorer than crops that are drilled, but the comparison is not a fair one; many of the former would never have been sown if the farmer had decided to wait for conditions which would have enabled him to prepare the field for the drill. It can certainly be stated that, though ploughed-in crops are sometimes failures, they are often successful; some farmers adopt the method as a general rule and follow it when time is no particular object.

When ploughed in, beans are sown in every second or third furrow. It is very desirable to harrow the field after the ploughing has been completed, because hoes will not run well along plough furrows, and if nothing is done the surface will be terribly rough for the binder at harvest. If the harrowing is done immediately after the ploughing the crests of the furrow slices will not crumble down well, and sometimes it is decided to postpone the harrowing till early spring; the danger then is that a late wet spring may make it impossible to harrow until the beans have started growing up, and then the operation may damage the plants seriously.

It is regrettable but true that year by year the bean crop is losing in popularity. Farmers are fully aware of its merits, as benefiting the soil and as providing protein-rich concentrate for livestock, but it always has been an uncertain crop and now that the old care cannot be lavished on it, yields are often miserably low. Hoeing, in and between the rows, is now often omitted and this makes the crop very foul if drilling distances are wide. The crop is therefore sometimes sown, as for cereals, with all coulters in the drill and all idea of hoeing abandoned; successful crops have been grown in this way but podding is rarely good and threshing results are often disappointing. Another method is to broadcast 4 bushels of seed an acre on the unploughed land and to plough them in with a fleet furrow, harrowing severely as soon as possible in spring; this last removes a lot of weeds and there is some evidence that it checks the clover stem-rot fungus which attacks the roots of beans. The repeated harrowing causes many casualties among the bean plants but that is allowed for in the heavy seed rate, which provides about double the required number of plants. Something may be done by improving strains of beans to make the crop more profitable, but few farmers will at the present time undertake, or risk, a large acreage of beans; probably the best advice to farmers is to grow only a small acreage and to do those acres really well or, alternatively, to grow beans in mixture with oats.

(4) DRILLING

Numerous efforts were made in very early days to produce a machine which would sow seeds in the ground, but it was Jethro Tull who first succeeded in perfecting an implement for the purpose in the early days of the eighteenth century; his name is indissolubly linked with drilling, and with horse hoeing which was thereby rendered possible. At the present time there are many different makes of drill on the market, and it is not proposed to enter here into the details of their construction, for which the reader is referred to text-books on implements. But there are points about drills which fall rightly into the purview of crop husbandry, and it is to them that attention will be devoted.

Some drills have fore-carriages for steering purposes, and though a fore-carriage necessitates an extra man in working and also slows down the work, because more time is taken in turning at the headlands, some farmers still prefer to have them. Certainly good work is often done with drills not fitted with fore-carriages, but in general their presence does help quite appreciably in keeping the drill straight, and preventing overlapping or gaps between drill widths; probably the value of straight work is aesthetic, rather than utilitarian, but it must be remembered that the drill rows are plainly visible for all to see for a considerable time, and neighbours are not slow to make rude remarks over a bad piece of drilling. This is a question that touches a farmer's proper pride, about which other people are wont to sneer, albeit that in their bones they envy it. The desirability of a fore-carriage is therefore a matter of aesthetics and such considerations carry much less weight than formerly, so that most new drills have no fore-carriage.

The two common forms of delivery mechanism are the cup feed and the force feed. The former does not give a favourable first impression, but nevertheless succeeds in achieving a uniform delivery; the grain reels about uncertainly, and the wonder is that such a constant amount finds its way into the coulter tubes. Since different sized cups are available—a drill being issued with two spindles, the functioning cups on which are changed when the spindles are reversed—and the speed of revolution is adjustable by changing a cog wheel, the cup feed delivery is very adaptable, and most drills of that type will sow the seed of any crop in any reasonable amount. The force feed is more limited in its range and is also apt to crush the seed, but has the advantage that its action is unaffected by rough tilths or sloping ground, and also that the rate of delivery can be changed very quickly. Comparative tests have failed to show any definite difference in the evenness of sowing achieved by the two methods, and it seems certain that with either the work is good enough for all practical purposes.

The coulter tubes may be telescopic or of flexible tubing and, both have proved themselves capable of giving good service. The plain V-shaped coulter has been most common in this country though the disk coulter has become more popular in recent years. It cannot be denied that the power of the disk coulter to cut the seed into the ground when the conditions are bad is important, but it is fair to urge that this power may tempt a man to use them when the seedbed is not sufficiently prepared. The spring type of press is easy to work, though in some drills it is impossible to take that type of press off the coulters entirely, and this may mean too deep sowing if the ground is very loose; the bar form of press holds the coulters in on even the roughest seedbed, whilst it carries an alternative for better conditions in the weights on the coulter arms.

The Suffolk type of drill embodies a fore-carriage, cup delivery, telescopic tubes, plain coulters and bar and weights for press. It is a heavy implement and normally requires three (sometimes four) horses on heavy land, though two suffice for the shallow sowing of root crops on a fine seedbed; for tractor work the fore-carriage is generally removed. The implement is solidly constructed and with reasonable treatment will do good work for very many years; its general utility has made it a popular type with farmers, many of whom are loud in its praise. The Bedford drill is constructed much more lightly, and though it has a fore-carriage this is controlled from behind the drill, and consequently one man less is required; otherwise it resembles the Suffolk type, but it is not suited to heavy land, and is relatively rarely seen. The American type of drill has a force feed, flexible steel tubing, disk coulters, a spring press and no fore-carriage; in many ways this type of drill is more efficient than the above types, but its disk coulters and light construction are against it in the eyes of many farmers.

A common width of horse drill is 7 ft. 6 in., though some are only 6 ft. 6 in.; the modern tendency however, is for drills to be wider so that the ground may be covered more quickly, and for tractor work widths of 20 ft. and more are becoming popular.

There are many special drills on the market. The combine drill which sows both seed and artificial manure has been mentioned in Chapter II; the commonest types of this sow seed and fertiliser down the same coulter tubes. Turnip drills are used for sowing small seeds on the ridge; they have bevelled rollers fitting the ridge, and one runs in front of each coulter to hold it in place. For some years strenuous efforts have been made to develop a spacing drill which will sow seeds singly at a predetermined distance apart in the row; a fair measure of success has been achieved and it is probable that spacing drills will come into general use for root crops. The most promising type has rubber bands, one for each row, that run between rollers in little runnels opened by coulters; the bands have holes in them and as each hole passes beneath a top roller a single seed is pressed through it. For a long time it has been the aim to produce a drill that delivers seed at ground level because of the loss of accuracy suffered as the seed falls down a coulter tube; these spacing drills do even better as they deliver seed below ground level and actually at the spot where it is required to lie. The type common at present is not a drill in the ordinary sense but drill units which are attached to the tool bar of a tractor; each unit has its own little hopper and is quite separate, four of them being normally used together. They were developed for sowing sugar beet but there are great opportunities for them with all root crops since they are flexible in use. Row distance is infinitely variable since it is determined by how the units are spaced along the tool bar; rubber belts with holes of different sizes and with various distances between the holes can be rapidly substituted for each other so that the units can deal with the seeds of any crop and can sow them at any required spacing within the rows. It will probably be found worth while with all crops to have seed graded by size and weight to obtain the best results with spacing drills. In districts of high rainfall these drills may revive interest in the turnip crop and spaced drilling may prove the ideal compromise between singled and unsingled kale. It might be thought that they will only function properly in ideal sowing conditions but this does not appear to be so; they have been used successfully on sloping and even on stony land provided, and this is important in all conditions, that the tractor be driven slowly.

ORGANISATION OF DRILLING

When a field is drilled sacks of seed are placed at intervals along one headland, and these should be spaced approximately correctly, so that as little time as possible is wasted in filling the drill; the writer knows of one farmer who, when drilling seed and fertiliser together, has one man doing nothing but filling these two into containers (e.g. bushel measures), so that when the drill reaches the headland these have only to be emptied into the appropriate boxes. The headlands should be drilled first because if left until the end they will be so trodden that the seed cannot be got into the ground; a good general rule is to start by going all round the field three or more times. With the

American type drill only one man is required, but he needs quiet horses if he is to do good work; with a tractor one man is required to drive the tractor and another to follow the drill (a platform is provided for him to stand on). With the Suffolk drill two men are required, one for the steerage and one behind the drill. The duties of the latter are to keep an eye on the coulters and to see that all are delivering seed, to keep the coulters clear of clods and rubbish which may be pushed in front of them, and to put the implement out of, and into, work at the headland. If a drill is stopped in mid-field it will travel about a yard on restarting without seed reaching the ground, so that a blank area will occur; if a stop cannot be avoided a handful of seed should be cast down just in front of the coulters, because blank spaces yield no crop and provide a chance for weeds to grow. With clovers and grasses the field may be drilled twice; a common practice was to drill one way with the heavy seeds (clovers and timothy) and then at right angles with the others; this was based on the belief that the heavy seeds in a mixture will tend to fall to the bottom of the seed mass in the drill box, and be sown first, but it does not appear that this sorting out actually occurs. This double drilling appears better on paper than it is in practice, because very often only one series of rows appears subsequently; if the field has been drilled one way and wet weather supervenes the position is an awkward one, as to put the drill over the field again may kill a large proportion of the seeds sown in the first instance if they have chitted. Where clover and grass seeds are being sown in a cereal crop it is good practice to drill them at right angles to the cereal rows, so that the majority of the seed may be between the latter; it would be even better to drill in the same direction with coulters running between the cereal rows, but this is not easy to do and is rarely attempted.

If drilling is carried out in good conditions and plenty of mould is present the seed should be covered up immediately behind the coulters, by the soil crumb falling in from the sides of the coulter marks, but it is unwise to trust to the thin covering thus provided, and the drill is normally followed by a light seeds harrow; this works parallel with the drill (and is in fact often attached behind the drill), and when the field is all drilled it is usual to harrow again at right angles to the previous working. Cross-harrowing gives a better cover to the seed and is believed to hoodwink the birds into believing that the rows run in that direction. If the seedbed is very fine it is possible to dispense with harrowing, by using the rings supplied with some drills to run behind each coulter; these may give quite a good cover to the seed, but they are apt to get tangled up in turning at the headland, and this may lead to waste of time. If the seedbed is very rough the ordinary seeds harrow may not give sufficient cover, and even a drag harrow may be used to follow the drill; this implement may also be used to move the land well with the object of killing weeds such as slender foxtail. If the harrowing precedes the drilling wet soil may be exposed which will become pasted on the drill, so that where it is necessary to use a drag harrow and a wet spell has been experienced, the drag harrow may be used as a covering implement; it is surprising how little of the seed is moved out of the coulter row even when the field is drag harrowed twice after drilling. Where the previous crop was kale, the old stems of which have been but poorly covered by the plough, only the lightest harrow must be used either before or after the drill, lest the stems be brought to the surface. Small seeds may be rolled instead of harrowed, with the object of pressing the soil firmly to them.

A final operation may be the drawing of water furrows, which is occasionally done on heavy clay fields with considerable slopes. By the time the seedbed has been worked down, and drilling and harrowing have occurred, the original plough furrows will have been largely smoothed out, though their locations will still be clearly visible. They are opened out again by taking the plough along them either once or twice and they may then serve a useful function as surface drains. But too often they are not connected to the ditch at the bottom of the field, and then they only serve to accentuate the undesirable tendency for the water from the higher parts to seep down and accumulate on the lower parts of the field. Water furrows, when carefully opened, may form a very efficient system of surface drains, and it is well worth while to lead them into the ditch, even though that has to be done with a spade.

BIRD MINDING

Several mentions have already been made of the depredations of birds, and of the fact that plant establishment may be seriously lowered, and the success of the crop jeopardised, by them. Seed is rarely treated to protect it from birds, though if effective the practice would have much to recommend it when seed is sown late in the autumn or very early in spring. Where birds are observed to be 'working' a field, some effort to keep them off should be made. The ordinary scarecrow is quite worthless, though a scare gun, which fires rounds at intervals, may deceive the birds for a time. For small areas stringing may be useful, but for larger areas this becomes impossible. A boy armed with a rattle is often employed, but in many cases a man may be working in a neighbouring field and he may be given a gun. Where pheasants are the culprits little can be done, but with rooks the best procedure is to shoot one, tear it to bits and leave the mangled remains on the field; the difficulty here is to get near enough to one to shoot it. A 'hide' is often built on the field for the purpose. With larks and starlings there is little that can be done, because when disturbed they only fly about 50 yd. and then they settle down to their fell work again.

TIME OF SOWING

This chapter would be incomplete without some mention of the times of the year when the various crops are sown. Table VIII represents an attempt to summarise information on this head. For the sake of incisiveness an optimum date of sowing is given for each crop, but the reader will realise that in practice only a small proportion of the crop is sown on a particular day. The actual date of drilling is very largely determined by the weather of the season, and must obviously await the arrival of the appropriate conditions of soil and climate; whether or not a farmer is well up with his work has much bearing on sowing date, whilst if he has a large acreage of one crop clearly he cannot sow it all on, or near, a particular date. Late-maturing varieties should be sown earlier than others, and on heavy land it is better to drill earlier in autumn than on light land, and it is usually impossible to drill so early in spring. Nevertheless, it is desirable to have some date in mind as the objective, whilst the next column of the table shows how widely round the date drilling is, in practice, spread; the common ranges given in the column are by no means extreme, but sowing outside these ranges produces generally, though not always, inferior crops.

If winter wheat is sown too early it may make much growth in a warm autumn and become what is usually termed 'winter proud'; this condition rarely leads to a good crop, since early lodging, with badly filled ears, is the usual sequel, whilst the crop is more liable to attack of rust and mildew. On the other hand, late sowing gives few early tillers and hence lessens the number of ears per plant; late sowing also results in slow establishment and physiologically weakened plants, because of the coldness of the ground about mid-winter, whilst the crop will be vulnerable to birds for a long period and may suffer very severely from their depredations. On rich warm land the main danger is from sowing too early, but on poor clay land the larger loss is occasioned by sowing too late. Where the soil is very poor and light, early sowing is favoured, because on such land the danger of lodging is negligible and it is advisable to obtain as much growth as possible before the advent of a summer drought. Spring wheat should be sown as early as possible, yields from April sowings rarely being satisfactory and in some cases of very late sowing the crop never comes into ear.

The general opinion is that winter barley should be sown in good time so that it may become well established before really cold weather sets in; there is considerable danger of winter killing if spring varieties are used. The idea that a crop will suffer less harm from frost when well established is a peculiar one which lacks any scientific backing; indeed, the contrary is probably more generally true, and certainly is so in the case of beans and peas. Some light-land farmers have, for many years, made a practice of drilling 'Christmas barley', that is, sowing during December or early January; though this has recently been hailed as a new discovery, it has long been known that there is little or no danger of winter killing of barley sown at that time. It is important that spring barley should be sown early,

TABLE VIII. SOWING DATES OF VARIOUS CROPS

Crop	Optimum sowing date	Common range of sowing dates	Notes
Wheat: Winter	October 20th	October 1st–December 10th	Later on good land
Wheat: Spring	February 15th	February 1st–April 15th	Special spring varieties
Barley: Winter	October 15th	October 1st–November 10th	Liable to winter killing
Barley: Spring	March 1st	January 15th–April 30th	May be sown about Christmas
Oats: Winter	October 10th	October 1st–November 10th	Special winter varieties
Oats: Spring	February 20th	January 15th–April 25th	—
Rye	September 10th	August 31st–November 1st	Occasionally in spring for feeding green
Beans: Winter	October 10th	October 1st–December 1st	—
Beans: Spring	February 15th	February 1st–April 1st	—
Peas	March 15th	February 15th–May 1st	Sown in succession for picking green
Tares: Winter	October 15th	October 1st–December 1st	Often mixed with oats
Tares: Spring	March 10th	February 15th–May 1st	Often mixed with oats
Lupins	April 20th	April 1st–May 15th	Sown in June or July for green manuring
Linseed	April 1st	March 15th–May 15th	—
Buckwheat	May 20th	May 10th–June 5th	Must not be above ground before last frost
Maize	May 20th	May 10th–June 5th	Must not be above ground before last frost
Swedes	June 5th	May 1st–July 1st	Earlier in northern than in southern counties
Yellow turnips	July 5th	June 1st–July 31st	Earlier in northern than in southern counties
White turnips	August 1st	July 1st–August 15th	Sown in September for 'turnip tops'
Kohlrabi	April 10th	March 20th–May 10th	—
Kales	—	March 10th–August 1st	Ready to feed about 5 months after sowing
Rape	—	April 15th–September 15th	Ready to feed about 3 months after sowing
Mustard	—	April 15th–August 20th	Ready to feed about 2 months, and to cut for seed 3–4 months after sowing
Mangolds	April 15th	April 1st–May 20th	—
Sugar beet	April 10th	March 20th–May 20th	—
Potatoes	April 10th	March 20th–May 20th	Earlies sown late February or early March
Short leys	April 1st	March 10th–May 1st	Sown under corn. Italian rye grass and crimson clover may be sown in September
Long leys and permanent grass	April 1st	March 10th–May 10th	May also be sown on bare ground or with rape, from July 1st to September 10th

there being much evidence that this leads to heavier yield and better quality. Crops of barley sown as late as the first week in May have proved satisfactory both in yield and quality, but these have been exceptions, and farmers are worried when drilling has to be postponed after the middle of March.

Ordinary oat varieties will not survive a winter but some are available which will, and these should be sown in good time in the autumn; farmers like to see them well established before the cold months (but see above) and, furthermore, this will advance the date of maturity, so that winter oats may be cut, and perhaps carted, before wheat is ripe. Spring oats should be in the ground as early as possible, principally because of the danger from the frit-fly, whose maggots eat out the shoots in May; the only defence against this attack is to get the oats well away in growth by mid-May. In the northern part of the country this pest is rare, and there oats may be sown late with impunity, and are, in fact, usually the crop to follow the latest folded fields of roots.

Rye which it is intended to harvest for grain is sown very early in the autumn, one reason for this being that this cereal tillers more in autumn than in spring, especially on the poor sandy soils on which it is grown; a further reason for early sowing is that then it is possible to obtain a valuable grazing of the crop before Christmas, after which the rye is allowed to grow a corn crop (it may even be grazed again in early spring and then be left for a corn crop). Winter beans should undoubtedly be sown early, so that they may be well established before shortage of other foods makes the rooks particularly devastating; it is also important that beans should be early so that they may set pods before black fly is liable to attack them in early June. On the other hand early sowing renders them more vulnerable to severe frost. Spring beans should be in the ground as early as weather permits in February; on heavy land, however, the possibility of drilling in February is problematical. The value per acre of the produce of beans is generally lower than that of wheat, and so beans often have to take their chance after all the wheat is sown; this procedure is a mistaken one, and is responsible for many of the wretched crops of beans that are seen. Field peas, also, should be sown in February if possible, despite the fact that when the ground is cold they are liable to rot; as has been seen earlier, dusting with an organic mercury compound will protect them. Blue peas—as the garden peas are called in agriculture—should not be sown until March, and it is common, whether they are to be sold direct as a vegetable or for canning, to sow several varieties in succession to extend the picking period. Tares are accommodating and can be sown over a fairly wide range of time with good prospect of success.

Lupins for seed should be in the ground by mid-May because otherwise if the summer is wet and cold the seed may not ripen. With linseed, also, sowing should be in April if possible, but it is important to clean the land before drilling because of the fact that a crop of linseed has very little smothering effect. Buckwheat and maize are both very sensitive to frost, and cold weather after the plants have come through the surface of the ground may be fatal to them; on the other hand, they need the full period of growth which can be allowed them, so that they are sown as early as possible, though few farmers will gamble on absence of frost until after the middle of May.

With swedes and yellow turnips sowing date becomes later as one progresses from north to south, because the main danger is that they may be too mature by the autumn; in that case they are very liable to attacks of mildew and they keep badly. Thus early May would be considered a suitable time for sowing swedes in Scotland and late June in southern England; yellow turnips mature more rapidly than swedes and are sown up to a month later than the latter. White turnips have been discussed in Chapter I, where it was seen that when the bulbs are wanted sowing should be in July or August, but that they may be sown after the corn harvest to produce winter-hardy tops. The kales may be sown over a long period. Now that they can be protected from the fly by seed dressing sowing dates can be varied according to when the crop will be required. If it is to be used to help out grazing in late summer and autumn the seed should be sown before the end of April but if it is wanted for winter feeding there is time for very useful cleaning of the land and sowing may be put off till late June in the southern counties or late May in the north. When mustard is wanted for seed it should be sown in early May, but it is usually sown for green manuring or folding at the end of summer; sowing should not be delayed beyond the middle of August as mustard is not frost hardy. Rape, on the other hand, is frost hardy and therefore its sowing may be as late as mid-September if it is not wanted for folding until late winter. Kohlrabi and mangolds are slow growers and so should be sown by the middle of May; if mangolds are sown too early many 'bolt' (that is, send up flowering shoots in the first year) so that they are rarely drilled before mid-April. Sugar beet are also liable to bolt if sown too early, but non-bolting varieties can be sown any time after March 20th; in order to spread the singling of the crop it is usual to sow sugar beet at intervals, but it has been shown conclusively that delay beyond the end of April definitely reduces yield of sugar. Potatoes which are through the ground are 'cut' by frost, and though this does not kill the plant it makes growth late and often reduces yield seriously. In a favourable spring many potatoes are planted in March, but April is the month when most are planted; often the work is not completed till well into May, but this shortens the period available for growth and usually results in a low yield. Early potatoes are often planted in February, but they are largely restricted to suitable districts, the chief characteristic of which is absence of late frost.

Short leys consist largely of red clover, and are invariably sown in spring under a corn crop, so that their slow establishment may occur whilst the nurse crop is growing. Young clover plants may be killed by very severe cold (they differ from most species in becoming more resistant as they get established), and sowing earlier than the middle of March is risky; on the other hand, by the end of April the nurse crop should become too high for the seeds drill to be used. The seed is sometimes broadcast into a corn crop which is 9 in. or more high, but then the young seedlings suffer severely from shading, whilst by late spring they are more liable to suffer from drought. Long leys and permanent grass are very often sown in a nurse crop in spring, the same considerations arising as with short leys. Permanent grass and long leys may, however, be sown on bare ground, in which case mid-April is a good time; some cleaning may then be done before drilling the seeds. From mid-May till mid-July is an unsuitable time for sowing pasture seeds, because of the danger that a blazing sun may cause the young seedlings to wilt and die; there is also more danger then than at any other time that the seed may germinate and then the plant die from lack of water, before its root system is established. Many fields have been sown down to grass very successfully after a summer fallow, seeding taking place from mid-July to early September; sowing later than that involves the serious risk that young clover plants may not survive a cold winter or may suffer severely from the attacks of slugs. For playing fields no clover is included in the mixture, and then very successful seedings have been made as late as the end of October.

It is typical of the uncertainties of agriculture that a direct and unqualified answer cannot be given even to this simple question as to when a particular crop should be sown. Farmers' opinions differ on the matter and, though some of the differences can be explained by variations of soil and district, many cannot; a man's personal experience always weighs heavily in the scale of his judgement, and many are undoubtedly misled by attributing successes or failures of crops to dates of sowing, when the real causes are to be sought elsewhere. Fortunately a beneficent Nature allows considerable latitude over sowing date, without exacting any ruinous retribution. During the last fifteen years there has been a distinct tendency for earlier sowing of spring crops and there is no doubt that this has contributed materially to the marked rise in yield which has occurred in that period. This gain has only been made, however, by the sacrifice of opportunity for cleaning the land by cultivation before drilling takes place.

CHAPTER VIII

AFTER-CULTIVATION

The word after-cultivation is used to designate all the work entailed in growing a crop, between the time the seed is covered and the time of harvesting. One important operation which may be carried out in the interval is spraying, either to kill weeds or to control blight in potatoes. The former has already been described, while for information on the latter the reader is referred to books devoted entirely to potatoes or to mycological works. It will be convenient to deal first with certain treatments to which wheat is subjected occasionally, and then to discuss after-cultivation in a more general manner.

Wheat which has become winter proud should have its growth checked during spring so that lodging, which normally follows this condition, may be avoided, or at least postponed until later in growth; the methods used for checking growth are grazing, cutting, and spraying with sulphuric acid. Where it is decided to run sheep or cattle over wheat the time for doing so must be chosen with care; a dry period is essential and the animals should be removed from the field if rain occurs, because there is risk of serious puddling of the surface. In some years a suitable period never presents itself during the spring, and then an intended grazing is best omitted. The usual month for this grazing is April, though a dry March may provide an earlier opportunity (very occasionally sheep are seen on wheat in November). Animals are not folded on wheat, but they are spread thinly over a good acreage and are generally removed as soon as they have eaten about half of the available growth. Rank patches of growth may be cut with a scythe, and in the fens it is common to run a grass mower over a wheat field during the first week of May; this is very drastic treatment, but it is the 'flag', rather than the young flowering shoots, which is removed. Spraying with sulphuric acid is more expensive but would appear less drastic and has the advantage of killing many weeds (in the fens valuable control of willow weed and hemp nettle results from this spraying); nevertheless, farmers who have adopted spraying are generally not enthusiastic over it, the opinion gaining ground that the crop is very slow to recover.

The practice of grazing winter proud wheat in spring is traditional but there is very little experimental evidence as to its effects; it is generally held to reduce lodging (this is, in fact, almost self-evident since the length of straw at harvest is certainly lessened) and to increase the yield of grain. In the past, sheep have been the livestock normally employed for the purpose and the primary aim has been to benefit the crop, rather than to provide food for the flock; latterly, cattle have frequently been used and the value of the food they get has been an important consideration. During and just after the War of 1939–45 farmers were often advised to treat wheat as a dual-purpose crop, to get it winter proud deliberately so that it might provide valuable grazing for cows in April and then a crop for grain; farming opinion, has, however, swung away from the practice and this is in accord with experimental results. These have shown that wheat in a normal state of growth (that is, not too forward) never gains from spring grazing and may suffer a severe fall in subsequent yield; to avoid any loss the grazing must be finished before mid-April and must be followed immediately by the application of a nitrogenous top dressing (around 0·40 cwt. of nitrogen). If it is expected that the grazing will be needed in the spring the wheat should be sown as a catch crop and should be ploughed up when the grazing is finished; there will be ample time to plant kale after it. Otherwise, wheat should only be spring grazed if favourable growing conditions have made it definitely winter proud, the object being to salvage a crop which is likely to suffer severely from early lodging; the grazing is, in fact, a corrective and it is improbable that there will subsequently be a bumper yield of grain. Winter oats may also be winter proud and they have been grazed in April, but it is by no means certain that they react to this treatment in the same way as does wheat; some farmers hold that spring grazing makes winter oats more, not less, prone to lodging.

Rolling, harrowing and hoeing are the common after-cultivations, and it is proposed to deal with these as generally as possible; it will be necessary to leave the potato crop for separate treatment at the end of the chapter, because in that case after-cultivation is very important and rather specialised.

ROLLING AND HARROWING

In after-cultivation these operations are practically confined to corn crops, of which wheat is the one on which they are most commonly performed; the case of winter wheat will be described in some detail, after which it will be possible to treat the others summarily.

A large proportion of the winter wheat grown in this country is rolled and harrowed in spring, except when that season is abnormally wet; in that case the land may never be dry enough to permit the work before the wheat has shot up too much to carry the implements without damage. The following are the objects of rolling wheat in spring:

(*a*) *To break a surface pan.* The extent to which a pan develops is determined by the fineness of the original seedbed, the amount, and the occurrence, of beating rain during the winter and, of course, by the proportion of clay in the soil; in some cases, on heavy land, the soil may run together into an almost continuous sheet which shrinks on drying with the formation of a network of cracks. Probably the chief harm wrought by the surface pan is that it will hold rainfall on the top of the ground (which, in itself, accentuates the panned condition) from which it will be evaporated; in addition, the aeration of the soil may be seriously impeded, whilst in extreme cases the wheat plants may suffer mechanical injury through the expansion and contraction of the pan on wetting and drying. The great majority of wheat fields are not in this condition in spring, but when it does occur a surface pan should be broken down. For this purpose a ring roll, used when the soil is dry, is remarkably efficient, the pan crumbling down beneath it.

(*b*) *To consolidate the land and pack the soil firmly round the roots of the wheat plants.* The hackneyed comparison in this connection is with the gardener, who 'heels in' his plants after the winter so that they may be firmly anchored, with their root hairs in close contact with the soil particles. But the roll does not resemble the gardener's heel, because the latter packs the soil tightly to a depth of 3 or 4 in., whilst the former's consolidating action only extends to 2 in. below the surface; this can be seen by observing how much below the rolled surface are the marks made by the horses' hoofs or tractor wheels. The benefit to be derived from rolling must clearly depend on the form of looseness of the soil at the time. Heavy land may be very loose in spring, so that feet sink into it to a depth of 4 in., and rolling that type of surface has no effect, beneficial or otherwise. On the other hand, land which is puffy on the surface may be rolled with great advantage. Light chalky land is very apt to get into this condition during a cold winter, the top inch or so of soil being heaved up, so that some of the young wheat plants are almost thrown out of the ground; this puffiness at the surface is just what the roll is capable of correcting, and experiments have shown that useful increases in yield may then follow its use.

(*c*) *To check the wireworm.* At first sight it appears ridiculous to suppose that a wireworm could be seriously incommoded by the passage of a roll, but farming opinion is unanimous in believing this to be the case, and is supported by the fact that it is on loose soils that the ravages of the wireworm most commonly occur. In some cases the beneficial effect is obvious, because after ring rolling it may be possible to see the wireworm in considerable numbers in the marks made by the rings of the roll; if they can be persuaded to come to the surface in this way, birds, especially plovers, will account for very many of them.

(*d*) *To level the ground for the binder at harvest.* The clods left on the surface of the seedbed after drilling will have weathered thoroughly during the winter and will crush down readily under the roll, especially under a ring roll.

The objects of harrowing are:

(*a*) *To make the surface of the soil loose.* The value of the surface mulch was discussed in Chapter IV; though a satisfactory explanation of the fact is not at present forthcoming, there can be no doubt that it does prevent the soil from drying out to the same extent as where the surface is firm. The harrow may be used to break up a surface pan, but for this it is rarely as efficient as a ring roll, and the tines of a harrow may move parts of the pan laterally, with some damage to the young wheat plants.

(*b*) *To control weeds*. This is an effect the importance of which clearly depends on the weeds present. Sometimes a wheat field carries established plants of speedwell and chickweed, sprawling over its surface, and many of these may be pulled out by harrow tines; in some cases it is necessary to stop frequently for the purpose of removing these weeds from the tines. But the most useful cleaning effect is obtained by killing young seedling weeds which are just coming through the surface. The ideal time for the spring harrowing of wheat is when the foot, drawn sideways through the surface soil, exposes many white thread-like seedlings; such weeds will be killed in vast numbers by harrowing, because the seed reserves are nearly exhausted and the roots are not established, so that movement by the harrow tines is fatal to the little plants. Unfortunately the operation must be performed at just the right time to be successful, but, accurately timed, it is extremely efficient; in Norfolk it is sometimes referred to as the 'red-weed harrowing' because of the vast numbers of poppies which it may destroy. The ring roll is also surprisingly efficient in killing these young seedlings, as the slightest movement of the surface soil seems to be sufficient to encompass their destruction.

(*c*) *To thin out the crop*. Harrowing removes roughly 5 per cent of the wheat plants bodily from the soil, and tears off, in addition, some of the tillers from the plants which remain. This thinning action is rarely harmful, and if the crop is very thick it may be definitely beneficial.

(*d*) *To level the ground for the binder at harvest*. There is one action of both rolling and harrowing—levelling the ground—which admits of no argument. Harvesters are complicated and expensive machines and suffer much wear and tear if taken over a field which has not been touched since it was drilled, when it may have been in a very rough condition; rolling and harrowing cost very little and are often held to be justified on this ground alone. The elimination of weeds is another undoubtedly important object, although success in its achievement is very variable. As regards direct effect on the yield of the wheat crop there is room for considerable doubt. On heavy clay land the roll appears to have no effect—beneficial or otherwise

—even when the soil is loose in spring, which condition is held by farmers to demand the roll; on the other hand the harrow has, in experiments, occasionally produced a very slight increase in yield, the improvement in aeration, and possibly the reduction in number of tillers, being held responsible. In experiments on light land the harrow has proved ineffective (always excepting a possible destructive action on weeds) whilst the roll has given good results (up to 20 per cent increase in yield). In practice the two operations are often applied to winter wheat, and though it is easy to argue that one or both of them might be omitted according to conditions, their low expense must be borne in mind before condemning them.

The time for the spring cultivations of autumn-sown wheat must be carefully chosen. The land must be dry enough, and in judging this it is not sufficient merely to look at the surface; the soil should be fairly dry to a depth of 3 in., because otherwise horses' hoofs or tractor wheels will pack it, the soil running together under their pressure. The farmer's saying 'A March roll makes an April fool' is expressive of a sound principle, although in the southern and eastern parts of the country, where a dry March is frequently enjoyed, the latter part of that month often provides the right conditions. At some date in April the wheat shoots, and when it is more than 6 in. high rolling and harrowing bruise and break it, and it may be some weeks before it recovers. Thus there is a restricted time when these operations may be carried out and the usual limits are from mid-March till the end of April.

There is no doubt that the ring is better than the flat roll for the purpose, because it crushes clods better, moves the surface more and, apart from the actual ring marks, leaves the surface in a looser condition. The lightest harrow is the type of that implement most commonly employed, though heavier sets are sometimes used, especially where much weed is present or where it is required to thin the crop; there is some danger that heavier sets, especially if used twice, may move the land so much that vast numbers of charlock seeds may be induced to germinate. The usual order is the roll followed by the harrow, so that the field may be left with a loose surface mulch. Where, however, the original seedbed was very rough, the weathered clods on the surface should not be rolled down on top of the wheat plants, but should be scattered first by the harrow.

The same considerations arise with the other cereals when they are autumn-sown as with wheat. Winter oats are often very thick in the row, because seed rates are rarely reduced sufficiently to allow for the early sowing, and to compensate for the fact that the common winter varieties are small seeded; a further reason why winter oats are commonly sown thickly is that thereby there is some insurance against mortality, should the winter prove a hard one. Thus it is very common to harrow winter oats severely in spring to thin out the plants. On the other hand, oats suffer more harm from harrowing than other cereals, so that unless the plant is a thick one spring cultivations are often withheld. Spring-sown cereals may also be rolled and harrowed when they are up, but the practice is by no means so common as for autumn-sown cereals. Spring cereals will often only just be coming through in April and no cultivation should then be given, or the young seedling plants will die; nevertheless some farmers risk this killing and harrow barley just before it appears above ground, this practice being remarkably efficient in killing young charlock seedlings. When spring cereals are established sufficiently to bear the cultivations without harm they are growing rapidly, and pressure of other work may preclude the operation until the crop is too far advanced in growth. In any case the need for levelling the ground is very much less urgent, because spring seedbeds are generally finer than autumn ones.

Beans are rarely rolled because the heavy land on which they are chiefly grown is usually too wet until mid-March, and by then they are starting to grow up; rolling beans when they are more than about 2 in. high breaks a large proportion of them. Harrowing is also liable to damage beans seriously, and is best avoided after the end of March; light harrowing, however, often follows the first hoeing between the rows to shake out the weeds which have been undercut by the hoe blades. Peas are even more prone to damage than beans, and rolling and harrowing are rarely performed with them. Root crops are not rolled nor harrowed later than a few days after drilling, because as soon as germination has commenced the young plants are very vulnerable to either operation.

HOEING

The functions of hoeing are:

(*a*) *To kill weeds.* This is the main purpose of hoeing, and in most cases annual weeds, cut by the hoe blade just below the surface of the ground, die forthwith; if the hoe blade is not sharp so that the weeds are dragged out of the ground rather than cut, and if the weeds are immediately trampled upon in a wet season, they may succeed in re-establishing themselves, but the vast majority die. Perennial weeds will survive hoeing but only in a weakened state and repeated hoeing will kill many of them.

(*b*) *To make the surface loose.* Apart from the arguments already advanced in support of a surface mulch, it is very important to get this condition in the case of root crops, because otherwise subsequent hoeings will be ineffective in weed destruction, as the blades will not enter the ground; it is therefore of paramount importance with root crops to get a surface mulch as soon as possible, and to keep it through the summer.

(*c*) *To remove surplus plants.* Most root crops are singled and the hand hoe is the common tool used for the purpose.

There are two methods of hoeing, hand hoeing and what is best described as row-crop work. In the latter, blades are pulled through the ground between the crop rows at a depth of from ½ to 1 in.; this is much quicker and cheaper than hand hoeing which, however, is the only method of undercutting weeds growing very close to the crop rows or between the plants in the rows. Formerly the blades for row crop work were carried by the tines of a horse-drawn machine and this method is still fairly common, but the modern method is to use a light tractor for the work. In the early stages of mechanisation, implements like ordinary horse hoes were pulled by tractor, but more recently row-crop tractors, with adjustable wheel-spans and power-lifting tool-bars, have been developed. The common arrangement still is for the tool-bar to be at the rear of the tractor but there are obvious advantages in a front, or underslung, tool-bar, since the driver can then see the work he is doing. There will undoubtedly be further advances with row-crop tractors in the future but already it can be said that the quality of the work can be just as good as with horses, whilst speed of working is greater and cost per acre rather lower.

Horse hoes may be single or multiple row implements. With the former there are normally five blades (sometimes only three) on two arms pointing forwards in the shape of a V; by means of a screw it is possible to vary the width between the arms at the rear, so that the hoe can be set to work the full interrow width. Multiple row hoes usually have some form of front steerage, on which the body carrying the blades can be swivelled to a certain extent; the path of the horse holds the steerage approximately correctly, and the man at the rear can control the path sufficiently to ensure that the blades run between the crop rows. It is important to deal, at one passage, with the crop rows sown on a particular drill width, because all of these are parallel and, if crooked, will 'wobble' together; the deviations of the drill can be followed by the man controlling the horse hoe, whereas if he is dealing with two drill widths he must inevitably cut up the plants in some rows. It is, therefore, correct to take a half or a whole drill width at once, and if a start be made on the outside drill width, the concord will be maintained throughout. With cereals (which are now very rarely hoed) only one blade works between each two crop rows and there may be six or seven, or twelve to fourteen blades in the horse hoe. With root crops there are usually three blades working in an interrow space, and a common number of rows, both with the drill and with the horse hoe, is four; the horse hoe does not cover, at once, four full interrow spaces, but three full ones, with half of each space at the side. If four full interrow spaces were taken the position would be as follows:

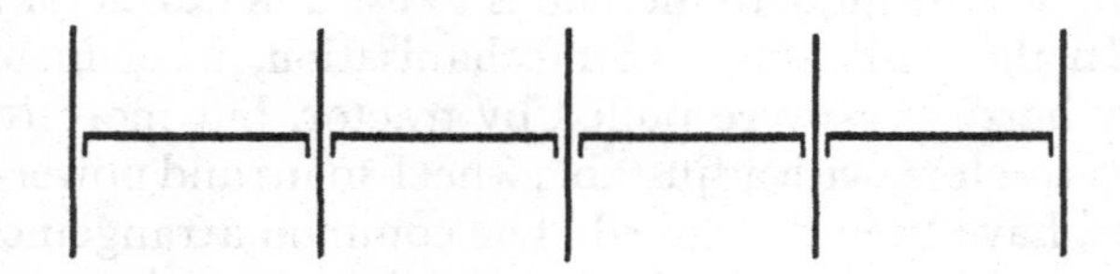

In that case the hoe blades would run right up to five crop rows, which must include rows sown at different passages of the drill; furthermore, the mid-point of the hoe, in front of which the horse walks, would be just over a drill row. The proper arrangement is as follows:

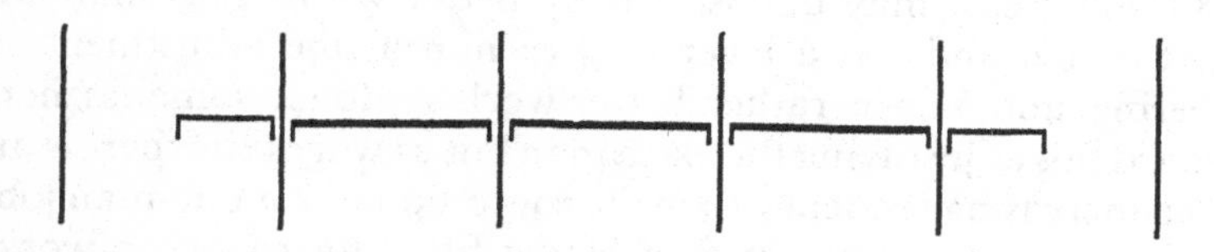

In this case the hoe blades only run close to four rows, which can be all in one drill width, and the horse walks in an interrow space. Similarly, if the horse hoe is only covering half a drill width it should cover one full interrow space and two halves, working twice across the field on the same drill width. It will readily be appreciated that exactly the same considerations apply to tractor hoeing, the wheels of the tractor normally straddling two crop rows.

The usual arrangement is to have two bars, one a foot or two in front of the other, to carry the tines; where three blades are working in an interrow space the centre one will be on the front bar, and the others on the rear bar, this staggering of the tines being very important in enabling surface rubbish to work through the hoe and leave it clear. There are many different patterns of horse hoe, into the details of which it is not proposed to enter. Low-built hoes are preferable in that they are easier to control, but they cannot be used late in the growth of a root crop, because then the bars will not clear the tops of the crop plants. When the crop plants are very small it is highly desirable that they should be protected from small clods rolling on to them. Some horse hoes are fitted with flat disks, one of which runs on each side of each crop row, whilst in others plates are provided for the same purpose; a farmer should possess at least one such implement for the earliest horse hoeings of root crops. The man controlling a multiple hoe is fully employed in avoiding cutting up the crop rows, and it is usual to give him a boy to lead the one horse which is required to pull the implement; having set the hoe properly, and working on drill widths, it is only necessary to watch one row of crop plants—to watch more than one is impossible, and only succeeds in producing a cross-eyed sensation. A row-crop tractor only requires one man, but skill and care are needed to do good work. Although a separate multiple row hoe for a tractor necessitates an extra man, the additional cost may be justified by better work. This man sits on the hoe and has a lever to give him a fine adjustment to steering and hence rather better work is done; some farmers regard this as justifying the extra man but as wages rise there is an ever-increasing tendency to make row crop work a one-man job.

There are various shapes of blades fitted for row-crop work. The narrow cultivator blades cut deeply into the ground, and are capable of cutting up a small piece of couch in process of establishing itself; they are useful for potatoes and possibly in the earliest hoeing of root crops, but are not to be recommended for later work, as they cut the little roots of the crop plants. The V-shaped blade is a common one and performs useful work, with moderate penetration. When the surface tilth is fine and the crop well up so that only shallow working is desirable, the L-shaped blade is commonly used; this is badly named, because the blade itself is merely a sharpened plate, the long arm of the L being provided by the vertical tine. A very common arrangement, with three blades working in each interrow space, is to have a V-shaped blade in the centre, flanked by two L-shaped ones. The general principle of working deeper in the early stages, and shallower as the crop gets up, and its roots spread into the spaces between the rows, is a sound one and should be noted. In all cases it is very important that the hoe blades should be sharp so that they cut cleanly through the weeds; it is also important that the soil should be dry enough, as otherwise it will stick to the blades and the hoe will not run constantly at the right depth.

Cereals were always hand hoed, and often horse hoed in the days when labour was cheap. At the present time hardly any cereals are hoed at all, and this is undoubtedly having a deleterious effect on the cleanliness of our arable land. Horse hoeing, even at present labour rates, is not very expensive, and a case might still be made for it; it can be done with normal distances between the rows of plants, but good work demands a surface reasonably level and free from loose rubbish, and rolling and harrowing may very reasonably precede horse hoeing. Hand hoeing of cereals is slow and expensive and is never done nowadays. Instead of the old type of hand hoeing, when the hoe was drawn through the soil and all the surface was moved, chop hoeing is not infrequently seen; this consists of walking through the crop chopping out specified weeds—very often thistles and docks—and a man may cover an acre or more in a day. April is the month for cereal hoeing, though it may be started in March, or be continued in May. Men are often seen chop hoeing later in the summer, whilst docks may be pulled out late in June or even in July.

Beans should certainly be hoed, though a wet spring may prevent it, in which case the doleful effect on cleanliness is only too patent in the following crop. The first row-crop hoeing is performed as soon as possible; it is not unknown for beans to be horse hoed before Christmas, but March is usually the earliest month in which the land is sufficiently dry. Single-row horse hoes or multiple-row implements can be used; this first hoeing may be followed by a light harrow, to shake out the weeds. If weather and pressure of work permit, a further row-crop hoeing should be given in late April; this can often be done in May, but some time in the early part of that month autumn-sown beans join across the rows, and then a horse or tractor will break many of them. Many farmers are well content if they can row-crop beans twice, but to complete the work, and to obtain the full cleaning effect of the bean crop, it is good practice to hand hoe them as well; hand hoeing is now rare and should always follow horse hoeing so that the men do not waste their time cutting out weeds between the rows. The wide drill rows of the bean crop provide a great opportunity for weeds in early spring, so that full advantage should be taken of any opportunity for hoeing; if beans are kept clean in the early stages the smothering effect later in growth will complete a very useful cleaning effect. The reader is referred to p. 234 where cultural methods designed to avoid hoeing of beans have been discussed. It should be realised that these methods have been forced on farmers by the rise in labour costs; few would argue that the elimination of hoeing is good, though many regard it as inevitable.

Peas also provide a good opportunity for hoeing, and it is probably even more important with them than with beans to make the fullest use of the chance; during summer they will lie flat on the ground and will not be thick enough to shade out weeds. If early hoeings are scamped, weeds will grow through the crop in summer, and many will ripen their seed before the peas are harvested. Two row-crop hoeings should be carried out in April or early May, and one hand hoeing is desirable though very rarely given. If these hoeings are omitted the crop will leave the field in a very foul state.

Root crops entail much more hoeing than corn crops, and in their case it is quite imperative that the work should be done thoroughly; if not properly hoed they are fouling, not cleaning, crops, and in bad cases they may be dominated by weeds, and fail miserably. The work necessary varies somewhat between the different root crops, and it is proposed to deal first with sugar beet which, being a cash crop, is usually treated well in this respect. The following are the operations normally performed:

(*a*) *Row-crop work to 'mark out'*. This consists of using a horse or tractor hoe with only one V-shaped blade running in the centre of each interrow space, of which it only moves a small proportion. The important point is that this should be done at the earliest possible moment, as soon as the rows are at all visible. If the roll followed the drill, the surface may be rather hard, and this hoeing will serve to break up a pan, which may be beginning to form, and it will thereby greatly facilitate the thorough hoeing which follows almost immediately; this early hoeing will also serve a useful purpose in 'marking out' the rows, which it will make much more easily discernible. There are many cases where hoeing is prevented by wet weather at the time the young plants are coming through, and then this initial operation may be omitted. Where beet are grown on the ridge a preliminary marking out is unnecessary, and thorough hoeing may be commenced as soon after drilling as any weeds appear; this is the main advantage of growing root crops on the ridge (see p. 145).

(*b*) *Thorough row-crop work with disks or plates to protect the young crop plants.* In this case the hoe blades are set to cover all the interrow space, and they should move the surface to within about an inch of the rows.

(*c*) *Singling and side hoeing.* This is the most expensive after-cultivation of sugar beet, and is often paid by piece work, rates running from 120*s.* to 180*s.* per acre. It cannot well be done before the young beet are in the 'four-leaf stage', because otherwise the plants will be too small to handle, and also because more may appear, to necessitate a repetition of singling. On the other hand, it is imperative that it should not be postponed more than a week or two after the beet are each showing four leaves, because of the greater disturbance the selected beet will suffer; numerous experiments have established the fact beyond all cavil that delay in singling seriously lowers the ultimate yield. But singling is very slow work, and therefore the general practice is to sow the sugar beet acreage of a farm in two, three or four separate sowings; in this way areas come to the singling stage in succession, and all can be dealt with at approximately the right period of growth. In singling, the usual method is for the man to face across the row, and to use an ordinary long-handled hoe; with this he strikes out most of the unwanted plants, though he may have occasionally to remove with his fingers those close to a plant he elects to leave. Another method is for men—in fact often women—to advance along a row on hands and knees, singling with a short-handled hoe and with the fingers. It is important that the beet which are left should not be disturbed more than necessary, and generally sturdy plants are selected for leaving; experiments have shown that special efforts to select only the very best beet, and to press the soil round them so that they do not fall over, are not repaid by any increase in ultimate yield. Weeds growing close to, and in, the row are of course removed, this part of the operation being referred to as side hoeing.

Singling is slow and very expensive and considerable ingenuity has been displayed in trying to reduce the labour of it, the ultimate object being to eliminate hand work altogether. Spacing drills hold out only limited hope because even with rubbed seed two plants are likely to come from some of the clusters sown and uncertainty of germination necessitates sowing the clusters more thickly than the plants are required in the final crop. To sow the seed in a small seedbed and then to transplant has little to recommend it as the saving of labour is problematical and the roots produced are usually very fangy. For some years hopes were pinned on 'cross-blocking', that is, cutting out surplus plants by doing row-crop work across the rows. Many different methods have been tried. Great care was taken to get continuous and regular rows of seedling beet by drilling on a very fine and smooth seed bed and opinion favoured drilling in rows about 20 in. apart and cross hoeing to leave plants 10–12 in. apart in the rows. Much was made of the fact that after cross blocking subsequent row-crop work could be done in two directions at right angles to each other and the wide drilled rows permitted work in one direction quite late in the growth of the crop. Cross-blocking, however, is going out of favour and is being replaced by down-the-row thinners; these have blades whirling round in a vertical plane and at right angles to the rows cutting out surplus plants. The speed of rotation relative to forward speed, the length of the little blades and the interval between them are so arranged that short lengths of the row, every few inches, are undisturbed. At first sight the operation is terrifying since it appears that the whole crop is being massacred but enough plants survive and the implement has even been used twice over on the same crop; in that case, statistical calculation will tell how many little bunches of plants will be left per unit of row length, but the calculation assumes perfect original spacing and the final inter-bunch spacing is variable.

Down-the-row thinners and cross-blocking have the important advantage that ground is covered quickly and although they leave more than one plant at each site they greatly reduce competition between the plants; thus after either operation the final singling may be delayed for two or three weeks without much harmful effect on crop yield and if a farmer has a large acreage of beet this is an important gain. It is generally claimed that either operation will speed up hand singling by 10–15 per cent but experienced hoemen are not very fond of dealing with the little bunches of plants that are left; their work has to be done more with fingers than with their hoes. Rising costs of labour are forcing farmers to experiment with these new methods but it should be realised that they entail some loss; a good hoeman can make a fairly regular crop out of gappy rows by carefully selecting the plants he leaves and this discrimination cannot be exercised by a machine.

(*d*) *Further row-crop work.* After singling, the plants must be given a period in which to recover (this they do remarkably quickly) and generally the field may be left for two or three weeks.

After that row-crop work is very desirable, to relieve the beet of competition, and to keep the surface loose. Yet another horse or tractor hoeing is often allowed, but by the middle of July the beet should be joining across the rows and row-crop work will then tear off leaves and may injure the roots.

(*e*) *Hand hoeing.* It is important that men with hoes should go through the crop once more, and this work is usually done in July, when haytime is finished. The work follows the final row-crop work and the main object is to prevent the weeds growing in the crop rows from seeding; an important subsidiary object is to remove 'doubles', that is, to pull up one beet where the singlers have left two growing together.

All this work makes the growing of sugar beet a continual source of worry and expense to the farmer, but it cannot be avoided if a successful crop is to be grown. Many experiments have shown the great loss in yield suffered through inadequate hoeing. The operations outlined above should be sufficient, and there is some evidence that they should not be exceeded; if, in an excess of enthusiasm, the farmer carries out several more hoeings than the above he is likely to reduce, rather than to augment, the yield, because the later hoeings will cut many of the fibrous roots of the beet.

Other root crops receive rather less after-cultivation than sugar beet. Mangolds germinate slowly, so that singling is later than with sugar beet, and time does not usually permit much post-singling work; when the mangold roots get large, hoeing is, indeed, liable to be harmful because the roots bleed if cut. Turnips and swedes are sown later in the summer, when weeds do not grow so profusely, and consequently they require less hoeing; a typical list of operations would be: row-crop, single and side hoe, row-crop. Kale grows up quickly and is an excellent smother crop when established, so that hoeing need not be continued late in growth. Kale should be row-cropped as soon as possible, but in many cases it is not singled at all, or is only chopped out to bunches; the crop should be side hoed, however, after which one more row-cropping normally suffices.

AFTER-CULTIVATION OF POTATOES

After-cultivation is an important part of the growing of the potato crop, and, though expensive, the work is fundamentally important to success; it must be remembered that this is a high value crop which may easily repay considerable outlay. The following operations may, in most cases must, be carried out:

(*a*) *Deep stirring between the ridges.* This is supremely important and may be commenced immediately after the planting is finished. On large farms the implement commonly used is a three-row ridger, with tines substituted for the mould boards, but on small farms the single-row horse hoe is generally employed; narrow cultivator points are fitted, three or five tines are used, and two horses are required. The implement is set to run to a depth of 3 or 4 in. below the bottoms of the furrows, and wide enough to move the shoulders of the ridges. It is convenient to have the horses abreast, but that involves one of them walking on the soil which has just been stirred; sometimes every other furrow is first done with the horses abreast, and then the intermediary furrows, in which the horses have walked, are done with the horses in tandem. The operation has little cleaning effect, because if it is done immediately after planting weeds will have had no time to grow, but it performs three very necessary functions. First, the worked soil having been put into the ridges, there is firm unmoved ground at the bottom of the furrows, and water will be held there and evaporated if the stirring is omitted. Secondly, by breaking up the firm soil in the furrows the operation encourages the spread of the weak-growing roots of the potatoes. Thirdly, the stirring provides more loose soil, and this will be needed later to build massive ridges capable of containing a good crop of tubers. If work is well in hand the operation may be repeated.

(*b*) *Rolling the ridges.* The reason generally advanced for this is that it brings up the moisture, but it has been seen that this movement of moisture can rarely be induced by rolling. In a normal spring potato ridges are usually not rolled, but in a dry spring a large number of fields are rolled a few days after planting; the operation probably helps in moisture conservation, by reducing the air spaces in the ridges, and so tending to check evaporation from the centre of the ridges. This is one of the very few operations for which the flat roll is preferable to the ring roll, because the latter tends to crumble away the sides of the ridges.

(*c*) *Harrowing the ridges.* This should be done just before the potatoes come through, the main object being to kill seedling weeds; thus it will be done some three weeks after planting and may annihilate myriads of weeds growing on the tops of the ridges, and thereby postpone or even take the place of hand hoeing. A further function is to bring the shoots through, as 'leggy' plants rarely yield well. Most farmers will stop the harrowing if many shoots are broken by the tines, but others do not object to this at all; naturally, the operation will not be performed if the potatoes are early, and frosts are expected. The harrowing follows the line of the ridges, and a light seeds implement is usually employed; an admirable implement for the purpose is a saddleback harrow, as that moves more of the outside of the ridges, whilst its tines do not sink so deeply into the tops of the ridges. If the field is rather foul very good cleaning can be effected by harrowing down the ridges and earthing up again two or even three times between planting and the emergence of the plants. Even perennial weeds may be controlled in this way since they get only a poor root hold in the loose ridges between a succession of disturbances.

(*d*) *Row-crop work.* It is necessary to row-crop at least once after the potatoes have come through. Horse or tractor hoes are used three or four times during the growth of the crop, the first being for the deep stirring mentioned above; the second time, the depth of stirring should be somewhat less than at the first, and on subsequent passages the depth is only about 2 in., the operation being then normal row-crop work.

(*e*) *Hand hoeing.* In order to cut the weeds growing between the plants on the ridges the field should be hand hoed once; the work is fairly quick as only a small proportion of the field has to be moved, and a man should cover half an acre or more in a day. One of the advantages of the harrowing referred to above is that it makes it possible to defer this hand hoeing until sugar beet singling has been completed. It is usually possible to dispense with hand hoeing altogether if a light harrow is worked along the ridges when the plants are well through. The action is drastic and some stems are broken off but the plants recover surprisingly quickly and the little weeds between the plants on the summits of the ridges are all killed. A farmer is well advised to keep away from the field whilst the harrowing is being done because it looks as if the crop is being ruined and his natural instinct would be to stop the work.

(*f*) *Ridging up*. All the operations mentioned will reduce the ridges, and by July the field may be almost flat. The double-mould board plough is sometimes used early in growth, either to cover the plants if they are just emerging in frosty weather, or to support the haulm as soon as it is 9 or 10 in. high. Normally, ridging up is done after all other after-cultivation has been completed, the date being July or early August. The objects of it are to support the haulm, to cover the tubers and so prevent them from going green, and to protect the tubers from disease spores falling from the leaves; ridging up is also valuable as a cleaning operation. The aim is to build a broad flat-topped ridge, and on loose soils, wet weather is preferred for the work, so that the soil may be plastered to the sides of the ridges.

This lengthy list of after-cultivations applies to maincrop potatoes; in the case of early potatoes the workings are spread over a shorter period, and their amount may be reduced slightly as then there is less time for weeds to grow.

CHAPTER IX

THE CORN HARVEST

For the gathering of the corn harvest it was customary to engage extra labour and to pay special wages for long daily hours, in order to cope with the work in reasonable time and to take advantage of favourable weather. The operations of harvesting have, however, proved very susceptible to mechanisation so that a farmer who is fully equipped may require no more than one man for every 100 acres of corn he has to harvest. The degree of mechanisation to be found on British farms is extremely variable and this is no cause for wonder in view of the varying conditions of soil, climate and topography which result in extremely diverse farming systems. The corn harvest on a large arable farm in East Anglia is a very different affair from that on a small farm, mostly grass, in Wales; on the former there may be 500 acres of wheat and barley to be dealt with and on the latter 5 acres of oats. It will be necessary to describe methods which many would regard as old-fashioned but which are still used for roughly half the corn crops grown in Britain. These methods, as applied to cereal and pulse crops, will be the subject of this chapter and methods involving the use of the combine harvester will be deferred for consideration in Chapter x.

The actual date of harvesting these crops varies very much. In most seasons farmers in the southern counties have finished their harvest before those in the northern counties have well commenced, whilst the date of starting, and, more still, of finishing, is greatly dependent upon the weather of the year. On light soils crops ripen from 1 to 3 weeks earlier than on the heavier soils, and further factors affecting time of maturity are the varieties and the sowing dates of the crops. It is, then, impossible to speak with any precision of the date of harvest. In the southern part of the country it usually commences about the last week in July, or the first in August, and it should finish by the first week in September; bad weather may so delay operations, however, that corn is still out in October, whilst in the northern part of the country it is not at all uncommon for the harvest to last into November.

STAGES OF MATURITY AT WHICH CROPS ARE CUT

During the early stages of maturation material is translocated to the grain from other parts of the plant, this process clearly resulting in some gain in weight of the grain; during the later stages the weight of grain tends to decrease somewhat by desiccation and this occurs either before cutting or after the corn is cut and in the stook. Thus there is a fairly long period (probably 10 days or a fortnight) over which the cutting date may be varied without appreciably affecting the final yield of grain. Corn is rarely cut so early that yield is lowered, except possibly in the case of oats, but sometimes cutting is deferred so long that considerable loss is occasioned by shelling, that is, by the grain falling out of the ear or pod and being lost. A piece of corn which is ripening earlier than other crops in the vicinity will attract very many birds (particularly sparrows), and the heavy loss occurring may decide the farmer to cut it earlier than he would otherwise. Where the grain is required for use as seed the crop should not be cut until it is fully ripe, because shrivelled grain not fully matured has a lowered germinative capacity.

As wheat ripens, the grain passes from a stage in which its contents may be squeezed out in a milky form, to one in which it has the consistency of cheese, and finally to a flinty hard condition. It has been shown that losses in yield of up to 20 per cent result from cutting the crop when it is in the milky stage. Some farmers prefer to cut wheat when it is still immature, because they believe that its quality is improved thereby, but this idea is fallacious, and, in any case, price is very slightly affected by quality. The correct time to cut wheat is when it is in the 'cheesy' condition; at that time the straw, which loses its green colour from the bottom upwards, will be yellow throughout its length. Final maturation, resulting in the flinty hard condition of the grain, will occur in the stook.

Barley should be cut when it is dead ripe, because the maltster wants it to germinate rapidly and evenly. As barley ripens the ears gradually bend over, and in some varieties they finally come to point directly toward the ground; the crop should not be cut before this bending is completed, and this will be a week or more after the straw has lost all traces of its green colour. The grain will be flinty hard, with transverse wrinkles beginning to form on its skin.

Oats are cut in a relatively immature state, partly because the straw has a considerable feeding value and becomes tough and fibrous if the crop is allowed to ripen fully, and partly because oat grains are liable to be blown out of the bells if the crop is mature. If cut too early the crop produces many shrivelled grain, and the usual stage is when the green colour of the straw has nearly gone, and before the glumes begin to fly in a wind; in the case of black oats cutting should be done as soon as the grain becomes black.

Rye is ready for cutting almost a fortnight before wheat. The straw loses its green colour much earlier than does that of other cereals, but yield is lowered if the grain is not firm before the crop is cut.

There is no well-known stage of maturity for cutting beans, which ripen from the bottom of the stalk upwards. A pod is regarded as ripe when a bean removed from it shows a black hilum (the scar on the bean at the point of its attachment to the pod); the usual recommendation is that beans should be cut when the pods are ripe half-way up the stem. At this stage the crop will still appear quite green because the upper parts of the plants will still be succulent; if, however, cutting is delayed until the crop appears black, great loss from shelling will occur. Practice varies much on the point, and one factor which precludes very early cutting is that constant stoppages of the binder are necessary if much green leaf is present; the feeding value of the straw is, of course, higher when cutting is done at an early stage, because there will be more leaf adhering to the stalk.

Peas should be cut as soon as the lowest pods are ripe, by which time the stems will be brown for about three-quarters of their length. Further delay is inadvisable because of the danger of shelling, of the depredations of birds, and of the fact that the straw has a relatively high feeding value if the crop is cut before it is dead ripe.

METHODS OF CUTTING

The sickle was used for cutting corn in the olden days but is much too slow for present-day conditions; it is sometimes seen, however, in use by smallholders, whilst the good sheaves it gives make it suitable for experimental plots. A man was expected to cut as much as half an acre a day by this method, whilst one man tying was expected to keep up with two sicklers. A fagging hook —essentially a small sickle on the end of a long handle—is not infrequently used for cutting peas. It is only about 100 years since the scythe displaced the sickle as an implement for cutting corn; a good worker can cut corn three or four times as fast with a scythe as with a sickle, but tying is slower work following the former, and the sheaves produced are untidy and ugly. The scythe is still a common implement in British agriculture. It is used by some smallholders who only grow a few acres of corn, and farmers are sometimes, and very reluctantly, compelled to employ it when corn is laid and twisted: often the lodging only occurs in patches, but in some years large acreages of the fens have had to be dealt with by this slow and expensive method. Where a field is opened out for a binder, that is, a swathe cut all round the outside of the crop to make a clear pathway for the machine, the scythe is used to cut this outside swathe; nowadays, however, few farmers open out for a binder. Very occasionally beans are pulled by hand. This is a slow and laborious procedure, but is sometimes deemed justifiable when the pods are borne very low on the stalks, so that a cutting machine would leave much corn in the stubble; this condition may arise through late sowing of the crop, or as a result of the field lying in a semi-waterlogged state for long periods. Peas are sometimes pulled up, and an ordinary horse rake, working to half its width, may be used for the purpose.

Where corn is badly laid, a grass mower is sometimes used for cutting; with this machine the cutter bar runs very close to the ground, and consequently it can get underneath corn which no other machine will cut. It does not clear a path for its next passage, so that there should be enough men tying to deal with one swathe ere the following one be cut; where the grass mower is necessary it is usually only possible to cut along one side of the field, and then five or six men are required to keep pace with the machine, the area covered being about 5 acres per day. There is something to be said for using the grass mower for cutting barley, the quality of which is improved if it is fully exposed to dews; in that case the cut corn is left to lie in the swathes and is carted loose, but the method is very rarely used. Peas are often cut with a grass mower, pea lifters being fitted to some of the fingers of the cutter bar; a swathe mover may also be fitted, and this clears a pathway for the next passage. The sail reaper has two considerable advantages over the grass mower for cutting corn; in the first place it clears a passage for the succeeding round, and in the second place it leaves the corn in sheaves ready for tying. This implement was used in the fens until quite recently and is still very occasionally seen in action. For laid corn it is preferable to the binder, in that it is much less liable to become choked with the twisted material. If a crop contains much green clover or weeds, it is undesirable to tie it up into sheaves until the succulent plants have wilted, and then a sail reaper may be used. This machine was also commonly used for barley in some districts famed for the production of high-quality grain; as noted above, quality in that crop is held to be improved by keeping the cut corn in an untied condition.

The history of the development of machines for cutting corn is an interesting subject, to which space will only permit a brief reference. A pusher type of implement, which consisted merely of a platform with sharpened teeth on its forward edge, is reputed to have been used in ancient Gaul, and a similar one was re-introduced in 1799 by Boyce; this cut off the ears and collected them. In 1808 Salman was responsible for a big step forward, when he introduced the reciprocating knife. In the next 30 years numerous efforts were made to perfect a reaping machine, and many are the claimants whose titles to the honour of inventing it have been urged. Machines which accomplished the work, more or less satisfactorily, were produced about

120 years ago, but it was not until 1865 that the sail reaper, as we know it to-day, emerged. From then progress was rapid, such details as canvasses to carry the cut material over the travelling wheel, packers and finally the knotter being evolved; it was in 1875 that Appleby incorporated all these in the first machine that cut the corn and tied it into sheaves. This binder achieved the object sought, and later progress has consisted in modification of constructional details, to produce the modern, very efficient, machine of which many models are on the market. By the end of the nineteenth century the binder was common on British farms. As the tractor has become commoner the power binder has found its way ever more widely into practice. Where the binder is drawn by horses, its working parts are necessarily worked by gearing from a broad travelling wheel, but in the power binder the drive is direct from the power-take-off of the tractor; this has several important advantages which the farmer has not been slow to appreciate. The drag of the working parts on the travelling wheel is liable to make it skid instead of revolve, and when the ground is soft from a recent wet period this occasions many stoppages; in the power binder the travelling wheel has no drag upon it, having merely to trundle the machine along. Where the corn is twisted and the crop is heavy and wet with dew or from a recent shower, the elevator canvasses often become choked, and then with a horse binder all the material on the platform must be taken off and tied by hand; with the power binder all that is necessary is to stop the machine's forward motion and clear it by working the power-take-off. Furthermore, if the crop is heavy, rate of progress may be slowed down without reducing the speed of the working parts, and this is another very useful capability of the power binder, which is lacking in the horse binder. Occasionally large stones get caught in the knife, and in the horse binder this brings the machine to a sudden stop; a power binder has a safety clutch at its attachment to the power-take-off, and this prevents such an obstruction from causing some catastrophic breakage.

New devices are being developed to make binders ever more efficient in dealing with laid and twisted corn. Crop-gatherers consist of arms which work from the reel and enter the crop ahead of the cutter bar with an action similar to that of a man lifting the corn with a fork; reels with spring, steel tines, often fitted to combine harvesters, are being adapted to the binder. Until recently the binder was normally used for standing crops and for those where lodging was not very bad; for worse cases farmers had to turn to implements which left the crop untied, the sail reaper or the grass mower. The time is rapidly approaching when these two implements will have no place in the harvest field, for the binder will be better able to cope with twisted corn than either of them.

CUTTING

It has been mentioned that one swathe all round the field may be cut with the scythe, before the binder commences work. Many farmers omit this and allow the machine and horses or tractor to crush down the corn next the hedge, cutting as much of it as possible later by sending the machine round in the opposite direction, when the outside sheaves have been stooked; the corn saved by cutting round with the scythe is certainly insufficient to pay for the operation, but to take horses and machines through standing corn quite naturally affronts most farmers' sense of what is right and proper. Normally the binder works round and round the field but if the corn is laid, one side (that on which the corn is leaning in the direction of travel) must be 'slipped'; in worse cases of lodging two or even three sides must be slipped, and where it is only possible to cut down one side, pathways for returning empty are made by moving sheaves. Often a supernumerary man is in the field to sharpen knives and to take over at meal times; he may also have to cut out small laid patches of corn with a scythe, or lift them with a fork so that the machine will cut them, and can always keep himself busy stooking. Where a corner of the standing crop is square the horses (or tractor) have to take a wide circuit, and in so doing they tread on some sheaves. It was the practice for the extra man to keep cutting the corners with a scythe, so that they were slanting and the horses need not walk away from the corn at all; nowadays that is very rarely done, most binders being fitted with a sheaf carrier, operated by the driver, to carry the sheaves delivered for about 15 yd. after a corner has been passed. In other countries the sheaf carrier is used continually, so that sheaves are dropped in groups, this being held to quicken stooking; but the practice has not found much favour in this country, although it is not unknown.

The heaviness of the work of cutting with a binder depends on the heaviness of the crop, whether the crop is standing or laid, the softness of the ground and the slopes in the field. Three horses are sufficient for the 5½–7 ft. machines with which they are used, but a team cannot be expected to work for more than about 5 hours. A low-powered tractor can cope with a binder and greater widths of cut are possible. Binders with 10 ft. cut have been made but they are clumsy in operation and cannot fit uneven ground. If the crop is a heavy one the canvasses and packers cannot cope with all the material and there are few binders in use with cuts of more than 7 ft. The area covered in a day varies much, but in good conditions up to an acre an hour is possible with horses, and double that rate with a tractor and a binder with a wide cut. Rate of cutting is greatly affected by the size of fields. If these are small there is much time lost in turning corners, whilst considerable time is wasted in changing the binder to pass from field to field; a binder in the working position is too wide to pass through a gateway, and it takes two men at least half an hour to alter it to the side-draught position in which it can be moved from field to field. Large fields make for rapid work, but when string runs out, or breakages occur, the stoppage always seem to happen at the corner of the field farthest from the place where string and spares are kept, so that much time is wasted in fetching them. Acreage covered is sometimes determined by sporting considerations; if most of a field is finished and only a few acres left overnight, nearly all the rabbits and hares will leave the crop during the hours of darkness. Some farmers therefore arrange that, when a field cannot be finished in one day at least 5 or 6 acres are left overnight; some admit frankly that their main object is to get some exciting sport, whilst others pretend to take the sterner view that vermin must be exterminated.

It must be realised that the binder is a complicated machine and that mechanical breakdowns may occur with maddening frequency; it is not only the waste time that galls the farmer, his main regret at stoppages being that full advantage is not taken of fine weather. The binder should always be thoroughly overhauled before harvest commences, and some spare parts and plenty of oil (grease for the more recent models) should be available. It is very important to have a supply of good string, as cheap string is continually breaking, and this not only causes stoppages but means that there will be many untied sheaves to hinder the work of stooking. A good crop requires roughly one 5½-lb. ball of string per acre. It is very important to keep the canvasses of the binder dry, as they shrink on wetting and the straps may be pulled off; when work is finished for the day the binder must be covered up to keep the canvasses dry, and the latter should also be slackened, for which a 'quick release' is fitted to most binders. The man riding on the binder must watch that the sheaves are being tied, operate the sheaf carrier, and adjust the position of the string on the sheaf, the height of the cutter bar and the height of the reel; if horses are being used he has also to drive them and is fairly fully occupied. In fine, sunny weather it is generally possible to start work by 8 a.m., but if the morning be dewy it is well to remember that the corn will dry quicker standing than when tied up in a sheaf; if it is intended to thresh the corn soon after harvest it is best to avoid cutting until the corn is dry.

Few farmers grow all crops, but where they are being grown the following is roughly the order of cutting:

(1) Rye, winter barley, winter oats, winter beans, peas.
(2) Winter wheat.
(3) Spring barley, spring oats.
(4) Spring wheat, spring beans.

STOOKING (OR SHOCKING)

Normally corn is stooked as soon after it is cut as possible, the exception being when there is much green ('seeds' or weeds) in the butts, in which case it is best to let the sheaves lie a day or two before stooking. The objects of stooking are to get the ears off the ground, to ensure that rain will run off and to expose the sheaves to the sun and wind; sheaves lying on the ground in wet weather get thoroughly soaked, are very slow to dry and, if wet weather continues, the grain is very liable to germinate in the ear. With wheat, from eight to twelve sheaves are put in each stook, but with oats no more than eight should be stooked together, because the straw is soft and wind will not draw through readily. With beans no more than six sheaves should be put in a stook, as the stalks are thick and slow to dry, and, these being sturdy, the stooks stand well. A common organisation is for three men to work round the field each taking two of the rows of sheaves left by the binder; the direction taken should be opposite to that taken by the binder, as otherwise it will be found necessary to turn round every time a sheaf is picked up. In some cases the stooks are put in parallel rows across the field, and this involves very little reduction in the speed of stooking, whilst it is rather more convenient when the stooks come to be carted. If this method is adopted it is possible to put the rows of stooks in lines running north and south so that both sides of each stook receive the same amount of sun rays; this is a point often considered important with barley, where evenness of sample is of paramount importance. This form of stooking has the further advantage that if wet weather delays carting it is possible to put heaps of farmyard manure between the rows of stooks or to cultivate the land. It is generally possible for a man to stook 4 or 5 acres in a day, so that from two to four men can keep pace with the binder. No satisfactory stooking machine has yet been evolved, though there has appeared an attachment for the binder which took the sheaves as they were delivered, and formed a stook, putting a string round it to hold it together; farmers were not, however, very impressed with the work done, and the fact that the stooks were left irregularly over the field militated against the machine.

If the crop is dead ripe, the weather set fair and there is no green in the butts of the sheaves, corn may not be stooked at all, the sheaves being picked up from the binder rows and carted within a day or two of cutting; this is fairly common with barley in a dry season. In windy weather stooks are liable to be blown over and then men must be sent over the field setting them up again. In wet districts stooks are often hooded, sheaves being laid along their tops, or opened and, as it were, clamped upside down over them; brown-paper stook covers are obtainable and these give good protection, but are expensive, very liable to tear and slow in use. A period of wet weather may necessitate moving stooks, the object being to turn each stook so that the inner side may be exposed to the wind; naturally this must only be done when the outer side is already dry, and when fine weather may reasonably be expected. If bad weather keeps the stooks in the field for a long time, and if the crop was undersown with seeds in the spring, the stooks must certainly be moved; it is remarkable how completely young grass and clover is killed where stooks stand for a long time. In a very wet season it may be impossible to get the stooks dry, and then the bonds of the sheaves may have to be cut; this is a final resort, only adopted by farmers with the greatest reluctance. Generally it is only the butts with which there is difficulty, and then, when the weather mends, all that is usually necessary is to pull over the stooks for a few hours before carting, so that the sun's rays may fall on the upturned butts.

FIELD ROOM

The term 'field room' is used to indicate the interval between cutting and carting a crop, that is, roughly, the time the corn stands in the stook. The amount of field room required by a crop depends on several factors. Ripeness at cutting time is one factor, since maturation is completed whilst the corn is in the stook. If it is intended to thresh the crop soon after harvest, it is important to allow sufficient field room so that maturation is certainly completed before carting; but even if it is intended to leave the stack unthreshed for some months carting cannot be advanced more than a day or two, because heating may occur in the stack. Lack of field room may be compensated to a certain extent by building small stacks through which air will pass fairly freely. Where there is much green material in the butts of the sheaves the stooks must be left until this is quite dead and brown before they are picked up. The main factor determining field room, however, is the weather; if that is hot and dry harvest proceeds apace, whereas if wet weather is encountered tempers are severely tried and harvest seems to drag on interminably.

At the time of carting, the grain should be hard and flinty, showing that maturation has been completed; the straw should be dry, and to test this the hand should be thrust below the bonds and into the butts, as it is in these places that drying is slowest. Dry sheaves make a rustling noise when moved, and for early threshing this condition is very desirable at the time of carting. When corn is carted too soon it is still 'sappy' and is liable to heat in the stack, which results in discoloured grain with, possibly, a lowered germinative capacity. The germination may also be affected deleteriously if corn is stacked whilst wet from rain or dew; in such cases there is difficulty in threshing if that operation follows soon after harvest, whereas if it is postponed the grain will be easy to thresh out, but the straw may be mouldy and dirty.

The required degree of dryness varies considerably between the different crops. In the case of wheat the straw is fairly strong and the stack will not settle down into a solid condition; consequently, if threshing is to be deferred for some time it is permissible to cart wheat when it is quite damp, and most farmers regard it as unnecessary to wait in the morning until the dew has dried. In fine weather wheat requires 7–10 days field room; it is a common saying that wheat should hear the church bells ring twice, that is, should stand in stook over two Sundays. Barley, being cut dead ripe, requires little field room and is often carted from 2 to 7 days after cutting; the crop is often undersown, and where there is much green in the sheaf butts more field room will be necessary. When barley suffers 'weathering' (that is, from rain) in the stook the grain acquires a dull, even a dirty, appearance, and this usually reduces its market value considerably; in catchy weather, therefore, barley is given priority and is picked up as soon as it is fit, irrespective of the claims of other crops. Oats usually require rather more field room than other cereals and are rarely carted earlier than 10 days after cutting; this is partly because they are generally cut whilst still somewhat green, and partly because the straw is soft and packs tightly in the stack, in which, therefore, little drying can occur. Country lore has it that oats should hear the church bells three times. Beans are given much field room because they are cut whilst still green and succulent, and their stems are thick and consequently take time to dry out; they are often cut first of all the crops on the farm and carted last, the interval being from 3 to 6 weeks. When beans have matured in the stook they are quite black, and the pods are very liable to open and shed the seed; moisture toughens the pods and so it is preferable to cart the crop when the dew is on it, or when a light rain is falling. The woody stems of beans make the stack very 'open' and it is possible to count on much drying in the stack; hence farmers often cart beans in a downpour of rain, and yet find little mould when the stack is threshed. Peas must be treated very carefully. They are not, of course, stooked, but are put into small cocks of a moderate forkful each. At intervals of 2 days or so these cocks are gently rolled over by men with forks, and are moved into rows between which carts can be led. The greenness of the crop at cutting necessitates field room of 10 days or more, to avoid heating in the stack, but when no sap remains the sooner they are carted the better; peas settle down very solidly in the stack and much mould results from carting them in a wet condition.

CARTING (OR LEADING)

It is now common to see motor lorries or sweeps in the harvest field, but horse-drawn carts or waggons, or tractor-drawn trailers, are the common vehicles used, these being organised to work as a set. The organisation varies from district to district and is necessarily dependent on the size of the farm. On small-holdings the common method is that expressively described as 'lob-cart', in which all the carts or trailers (usually only one or two) are taken to the field and loaded, and then brought together to the stack for emptying; the method is clearly suited to small gangs, and is not infrequently used on larger farms in carting hay from outlying meadows. On all farms of substantial size a set is organised in the same way as described in Chapter II for farmyard manure; at any one moment there will be one cart or trailer in the field being loaded, one at the stack being emptied, and one or more (determined by the distance to be traversed) in transit. One very efficient organisation is based on four-wheel trailers to which a tractor or a horse can be quickly hitched; a horse is used to move from stook to stook and a light tractor to run full trailers to the stack and to return with empty ones. In the field there are required one loader and one or two men pitching up the sheaves; if horses are used the latter should, if possible, include the horsekeeper who, if he is worth his salt, will see that loads are not too unbalanced, and that a load is completed at the top, rather than the bottom, of a hill. In some districts waggons are used and two men are allowed for loading the sheaves pitched by two other men; in other districts one diminutive lad is expected to load for two strong pitchers, but if the latter put the sheaves up handily this cannot be quoted as a gross case of the exploitation of child labour. When beans are carted it is customary to tack old sacks on to the fore- and hind-ladders of carts, to catch the grain that falls out of the pods; this procedure is also common with peas because the trampling of the loader causes the pods of pulse crops to open. Shelling also occurs with cereals, where the corn has stood for long in the stook, but the grain is shed as the sheaf is being pitched, and sometimes quite a shower of it falls to the ground as each sheaf is put on the cart. At the stack three or more men are employed, the number depending on the size of the stack. Of these workers one is responsible for the building of the stack and does the actual laying of the sheaves, the others passing them to him; the work is greatly lightened when an elevator is used to carry the sheaves to the top of the stack.

Round, oval and 'square' (really, rectangular) stacks are commonly seen in this country, and there seems little to choose between these shapes, the one selected being determined by local custom and the preference of the stacker. The size of stack, where the crop available is more than sufficient for one, is worthy of some consideration. It has already been said that corn that is not thoroughly dry when carted should be built into small stacks, through which air will pass fairly freely. In the wetter districts of the country this is the normal condition and consequently custom is rigid over the matter, small circular stacks of about 4 yd. diameter being the rule; as many as 100 of these little stacks are sometimes seen in one stackyard, and it is not very uncommon to build large triangular wooden frames, or faggots, into them to help aeration. But in dry districts larger stacks can be built with safety, and bottoms as large as 15 by 10 yd. are sometimes laid out; it must be remembered, however, that the larger the stack the more men will be required in the building of it, because of the greater distance the sheaves will have to be passed. It is convenient to have a stack that can be built, and later on threshed, in one day. This size is about 8 yd. diameter for a round stack, or 8 by 6 yd. for a rectangular one, and one of these dimensions will contain the produce of roughly 10 acres of a normal crop. Some thought should be devoted to the siting of stacks. It must be remembered that they are temporary, and that the straw ricks to which they will later give rise are more permanent; it should be arranged that straw ricks are so placed as to involve the minimum carriage of straw to the place where it is to be used, and a rick may well be put alongside a bullock yard for which, in fact, it may provide valuable shelter. In order to save time when threshing, corn stacks are often placed in pairs so that one setting of the threshing machine will suffice for two stacks; 4 yd. is the correct distance between two stacks when it is proposed to put the threshing machine between them. A stack must be in a dry position, and a stackyard should always be well drained. The stack bottom commonly consists of a layer of straw or hedge brushings; faggots raise the stack well off the ground but seem to encourage rats. Rats and mice take a large toll of the grain in a stack which stands a long time before it is threshed, and the loss is sometimes avoided by building a stack on steddles; these constitute a skeleton iron bottom raised some 2 ft. above the ground on iron or concrete pillars, the top of the pillar being in the form of an inverted saucer. Rats and mice find it impossible to negotiate these saucers, but it is important to leave nothing against the stack to provide them with a means of ingress; steddles are not very common in this country though the price of them is reasonable.

Competence in the building of a stack cannot be acquired from a book, and the reader who seeks it is urged to learn from experience, choosing for his first attempt a stack in an unfrequented place; few things are more depressing than a corn stack with large bulges and subsidences, propped up precariously by wooden beams. A few general statements on stack building are all that will be attempted here. The stacker works round and round the outside of the stack laying courses three or four sheaves thick; the outer sheaves (the butts of which must always be to the outside) are held in place by 'binders' laid nearer the centre of the stack. It is very important to avoid what is usually termed a 'duck's nest', that is, a stack in which the centre is lower than the outside; this must be avoided from the outset by starting with a stook set up in the centre and working outwards till the edge of the site is reached. If the centre of the stack is kept low during building, the sheaves at the outside will all tend to slope downwards and inwards, so that rain falling on one of the straws of a butt will run down that straw into the stack; if, however, the centre is kept high, sheaves will slope in the opposite direction, and rain falling on the butts of outside sheaves will drip off the ends of the butts on to the ground. This can be further ensured by building the stack with its sides sloping slightly outwards, but the tyro is warned of the danger of over accentuating this slope; as the stack settles there is a tendency for the sides to be squeezed out, and a pronounced initial bulge leads therefore to disaster. Where no elevator is used carts should be emptied roughly alternately on opposite sides of the stack, if that be possible, because otherwise disproportionate trampling will mean uneven settling of the stack; furthermore, with a normal-sized stack one cartload is usually sufficient for about half a circuit of a course of sheaves, and consequently emptying alternatively from one side and then the other means that the distance that the sheaves have to be thrown is reduced. When the sheaves are pitched on to the stack by hand the height of the stack walls cannot be much more than 12 ft.; a greater height is common where an elevator is employed. The angle of the roof is approximately 45°, and the importance of keeping the centre high during building is even greater in the roof than in the body of the stack; the first step, therefore, in making the roof is greatly to accentuate the extra height of the centre, so that the outside sheaves of the roof may slope upwards and inwards at an angle of some 30° to the horizontal. Where no elevator is used a 'pitch hole', or gap, must be left in the roof, a man being stationed there to take the sheaves from the man on the load; special iron frames are available to give the man in the pitch hole a foothold and prevent his slipping out. The last load put at the top of the stack is dealt with slowly because it is important to get as sharp a point, or ridge, at the top as possible; the stacker wants time to place each of the last sheaves carefully. With peas a sharp finish cannot be obtained and, in addition, the settling which rapidly occurs soon gives a flattened top; after a few days, therefore, some forkfuls of straw should be put on the top of the stack.

A gang engaged in carting requires careful organisation. The farmer should see that the gang is well balanced so that work proceeds steadily with everyone fully employed; two strong and active pitchers in the field and three or four old-age pensioners on the stack is not the sort of organisation which is to be commended. In the morning, and after meal times, empty carts should be sent out early to the field, so that full loads are quickly available at the stack; attention to details like this makes much difference to the amount of work done in a day. This varies greatly, factors affecting it being the standard, adequacy and balance of labour, hours worked, heaviness of crop and roughness of carting (if a rough cart track has to be traversed each load may have to be roped, and this means loss of time); a full and active gang should clear 10 acres a day of a normal crop of wheat, oats and peas, rather more of beans and up to 15 acres of barley. If weather is unsettled it is best to clear parts of fields near hedges, since it is there that drying will be slow if rain interrupts the work. In some parts of the country it is still the custom to leave one stook in the field, and antiquaries have often interested themselves in this, regarding it as a survival of an ancient custom of making votive offerings to the gods. The custom is, however, purely materialistic. It was the custom to permit women and children of the village to go into corn fields after they were cleared, to glean such ears as they could find, for feeding to their fowls or even for rubbing out and grinding the grain; most farmers are still willing to allow the gleaning, but now there are few women anxious to do it. The stook is left to show that the farmer has not finished gathering the crop, because it is usual to horse rake the field after carting; the odd stook is picked up with the draggings. These draggings are sometimes used to top the stacks, but more commonly the draggings from all crops are put into one separate stack, which, when threshed, yields mixed grain suitable for feeding to poultry or for grinding on the farm; alternatively, the draggings may be given unthreshed to pigs or poultry which will have no difficulty in finding the grain contained in them. Bean stubbles are often not horse raked, since a large proportion of the grain is shaken out of the pods and lost in the process; a good practice is to run a flock of ewes over the bean stubbles which they will clear up very satisfactorily.

THATCHING

Many farmers possess Dutch barns into which the corn is stacked as it is carted. Dutch barns ensure that the corn is kept dry and eliminate the necessity of covering the half-finished stacks; this is a great convenience, and to it must be added the fact that threshing can often be carried out under cover. On the other hand, Dutch barns bring the corn all together, and the uses of the straw may require different sites for the stacks; concentration of the material is also to be deprecated on account of the danger of fire. Undoubtedly Dutch barns are very useful, however, and it is only the question of cost which prevents their wider and more rapid spread. In their absence stacks must be thatched, unless they are to be threshed very shortly after harvest. A well-built stack will suffer very little harm from a moderate rainfall, but the possibility of a deluge makes early thatching desirable; this is usually the first work after harvest and may even be commenced on wet days during harvest. Except for the discomfort of the men, rain is desirable during thatching, as it toughens the straw.

It is usual to speak of thatching as a lost art, but some neat stack yards are to be seen about the country providing evidence that there are yet men available who are masters of the art. In thatching a stack the first thing to do is to fill up any irregularities in the roof with straw. The thatcher's mate is kept busy preparing the straw which is usually that of wheat, though rye straw is preferable if it is available. The straw is laid out as straight as possible in a large heap, over which water is thrown, and it is best to do this overnight, so that the wet straw may toughen; the straw is then drawn out by hand, the heap combed out, and bundles, or yelms, of 6–9 in. thickness, formed. These are carried up the ladder to the thatcher on a 'thatching jack', which resembles the bricklayer's hod. The thatcher lays the straw in a strip alongside his ladder, starting with a double thickness at the eaves, and working upwards, so that each straw leads water that falls on it out of the stack. When a strip is laid it is combed with a little rake, and held in position with spits (i.e. light wooden stakes) and string; a row of spits is put round the stack about 1 ft. above the eaves, and further rows at intervals of 1½–3 ft. It is very important that the thatch should project from the eaves as it is there that water is most apt to penetrate into a badly thatched stack. A man who takes a proper pride in his work trims off the straggling ends of straw with sheep shears, and, in some cases, trims the walls of the stack; this last is justifiable only on aesthetic grounds but is a small and harmless extravagance.

Payment for thatching is often by piece work, the unit being a 'square', which is 100 sq. ft. of roof; with the common size of stack in the southern part of the country, there are approximately 10 squares in the roof. The thickness of straw is about 4 in. and about 2 cwt. of straw are required per square. Piece-work rates of 9*s*. per square will therefore make the total cost (allowing £3 per ton for straw) about 15*s*. per square, that is, very approximately, per acre of crop thatched. This is the cost that is directly saved by a Dutch barn. With the big increase in corn acreage during the War of 1939–45, thatching became a serious problem and various methods of eliminating or quickening the work were tried. Sisalkraft paper, laid in over-lapping strips over the ridge of the stack and held down by a wide-meshed string net, will keep a stack dry for some weeks, but the paper is apt to tear or gape in a gale and the covering will not often see a stack safely through a winter. A machine for making mats of thatch has done useful work. The mats can be made during wet weather and stored for later use; much binder twine is used by the machine to stitch the straw into mats and storage requires some care lest mice nibble the string and destroy the stitches. The mats are laid horizontally along the stack roof, working upwards, of course, so that the overlap tends to run water out of the stack, and fixed by a large needle which stitches them to the stack. Where the work is well done the appearance is neat and protection is fairly good for some months; but the mats are thin and the covering is definitely inferior to well-laid thatch. Mat thatch is now rarely seen.

In the northern counties, where many small circular stacks are built, the procedure in thatching is very different. A sort of drill has been evolved, which requires one man on the ground, one on a short ladder and one on a long ladder; the straw is laid, and then ropes are thrown across the stack and fastened with spits, so that the roof is covered with rough network. It is held that this method is quick and that the thatch is firmly held against being blown off by a high wind. Three men are expected to thatch nine stacks a day by this method, but to a southern farmer this is not impressive, in view of the very small size of the stacks.

THRESHING

A considerable proportion of corn is threshed very shortly after harvest, but the threshing of the remainder is spread fairly evenly throughout the winter, a certain amount being left as late as the following spring and summer. The following are the factors which decide the date on which a stack is threshed:

(*a*) *When it is fit.* If the weather is dry at the time, corn may be fit to thresh direct from the stook and 'carrying and threshing' is commonly seen. The great argument in favour of threshing direct from the field is that it saves handling, since the building of the corn stack is eliminated, but against this must be set the fact that a well-organised gang can cart more quickly to a stack than to a threshing machine. Furthermore, delays may be occasioned by the weather, as damp corn cannot be threshed and, unless a stack is built of part of the crop to act as a reserve, heavy dews and showers cause wasted time. Most farmers hire the threshing set, and it is imperative to keep it fully employed while it is on the farm; this requires a larger staff of labour than is necessary for carting to a stack. It must be emphasised that the corn must be fully ripe if it is to be carted and threshed at the same time; with pulse crops this is rarely the case, as even after the generous amount of field room usually allowed them, the grain of beans and peas is still rather soft when the crop is carted. These crops thresh better, therefore, after they have been in a stack for some weeks, as also do cereals which are carted whilst at all immature or damp.

(*b*) *When the grain is required.* Seed corn for autumn sowing must clearly be threshed soon after harvest, whilst for spring sowing it is best left unthreshed until shortly before it will be required. In the case of barley, maltsters buy up a large part of their year's requirements at about the time of the Brewers' Exhibition in October, and prices generally rule highest at about that time of the year; there is, also, a belief that the quality of barley improves during the first 6 weeks that it is in the stack, and consequently producers of high-quality barley thresh most of that crop during October and November. As regards wheat, the guaranteed price rises by £5 a ton from September to the following summer. Beans are unsafe for feed and cannot be ground until 4 or 5 months after they have been stacked, maturation still proceeding slowly throughout this period. During the autumn months the market price of old beans is higher than that of new beans, and many farmers leave the crop in the stack a full year before threshing it; by this they gain in safety of food (and in convenience of feeding) rather than value, as old beans are much smaller than new ones. Oats, also, are regarded as unsafe for food before the Christmas following harvesting (the same is true of rye), so that the threshing of that crop is often delayed for some months.

(*c*) *When the straw is required.* Very often a farmer has little or no straw left at the end of winter, and then he must thresh some corn quickly after harvest, to provide him with material for thatching and for bedding of stock. Some farmers lay great stress on the superiority for feeding purposes of freshly threshed straw over that threshed for a long time; in parts of Scotland it is common for each farmer to own a small stationary thresher, with which he provides himself with fresh straw at approximately weekly intervals during the winter. Fresh straw undoubtedly smells sweeter than stale straw, and is certainly more palatable to stock, but that is often due to the careless way in which straw ricks are built, little trouble being taken to prevent rain from penetrating right to their centres; this does not appear, however, entirely to account for the superiority of freshly threshed straw, although no other reason for the phenomenon is known.

(*d*) *Damage suffered in the stack.* A badly built and badly thatched stack deteriorates seriously as time proceeds, and it is better to thresh it early than to leave it to spoil. Rats and mice may play havoc in a corn stack which may come, by the end of winter, to contain little more than straw and the husks of grain; the date of threshing many stacks is advanced to avoid this loss.

(*e*) *When cash is required.* In farming circles it is polite to assume that a neighbour threshes because he needs the straw; unhappily this is too often merely a pleasant fiction, his main object being to quieten his more pressing creditors.

In early days corn was threshed by the treading of oxen or slaves, and later by means of the flail, whose use still persists for small amounts of grain and, very occasionally, to avoid cracking seed beans, after an abnormally dry harvest. The old barns so commonly seen were mostly built for threshing, several men spending nearly the whole winter in them wielding the flail; it is easily understood how the labourers regarded the threshing machine as a robber of their winter employment, and how its introduction in the years following the Napoleonic wars led to agricultural riots. In recent years the same feeling again has been experienced, though fortunately without a like violence, about the effect of the combine harvester on employment.

Threshing sets, consisting of a thresher, elevator, chaff-cutter and straw-baler, together with the engine or tractor to work them and to pull them from place to place, are owned by contractors, who let them to farmers; many of the larger farmers own elevators and tractors, and there is a tendency for more of them to purchase threshers, so that they have a complete set of their own, and are independent. The usual hire charge for a set is about £10 a day and this includes two men who accompany the outfit; the charge is increased by some 30*s*. when a wire-baler is used, the farmer also having to pay for the baling wire. For a description of the threshing machine the reader is referred to text-books on agricultural machinery; a knowledge of its inward parts is, in fact, not essential to the farmer, since the men in charge are usually competent to make the necessary adjustments. The farmer has merely to judge the results of the threshing.

If he finds grain remaining in the straw he will insist that the concave be set closer to the drum. If he finds that the grain is crushed, he will take a sample from the little trap at the foot of the corn elevator; if the corn is broken there it is the concave which must be set more widely, if not, the chobber. The corn sample must also be examined to see that it is well graded and that the chaff has been removed; if the sample leaves something to be desired adjustments must be made to the rotary wire screen or to the awner and chobber. In the case of barley it is important to avoid skinning the ends of the grains, and some farmers insist that a short length of awn be left on the grain to prevent this. It is surprising how rarely self-feeders are used with threshing machines in this country; the common practice is to have a man cutting the bonds of the sheaves and another feeding the loose sheaf into the drum. A self-feeder eliminates the latter and in some cases is fitted with knives to eliminate the former; if the bonds are cut in this way, however, the straw is much cut up and the string is included in it, and occasionally this causes trouble if the straw is used for fodder. In feeding by hand the aim is to keep as even a rate as possible, each sheaf being opened out and spread over the width of the drum; where unbroken straw is wanted the sheaves should be fed in horizontally, and in some cases with rye the grain is 'buffed off'—that is, the ears are held to the beaters and then the sheaf withdrawn. When beans and, to a less extent, peas are being threshed, odd grains are carried round by the beaters of the drum, from which they emerge through the feeding opening at considerable speed; to be hit by one of these is, at best, uncomfortable and at worst, when the hit is in the eye, serious. The general practice in some parts of the country is to fix two hurdles over the drum in the form of an inverted V, and to cover the hurdles with sacks against which the flying grains impinge; in that case feeding is done from one of the open sides of the V, a fork being used, or a fagging hook, to cut the bonds as the sheaves are dropped into the drum.

The straw may be built into a loose rick, in which case an elevator is placed to take the straw as it comes from the shakers, the elevator being driven by a belt from the thresher; an alternative is to have a trusser to take the straw and tie it with string into loose bundles. Most farmers nowadays prefer the straw in tight, wired bales and for this the baler works, in place of elevator or trusser, direct from the thresher. Another possibility is to cut the straw up into chaff, for which purpose a chaff-cutter, again driven by a belt from the thresher, is placed in front of the latter; a man takes the straw as it falls from the straw-shakers of the thresher, and feeds it into the chaff-cutter, but it is difficult for him to keep pace with the threshing. The cavings consist of broken pieces of straw and of leaf and, if clean and sweet, they make good fodder; sometimes they are accumulated in a separate heap, but more generally they are put into the hopper of the elevator and incorporated with the straw. The chaff of cereals consists of the glumes, and in the case of wheat (except bearded wheat) is good fodder; with oats the chaff is apt to be dangerous food, as it gets into the animals' eyes; with barley the chaff consists principally of awns and is generally regarded as useless, but it has often been used successfully as fodder, when mixed with pulped swedes. Sometimes the chaff is taken to a safe place and burnt, but this is usually to the farmer's shame, since the common reason for the procedure is that the chaff contains many weed seeds.

The number of men in a full threshing gang is from 8 to 10, their distribution being as follows:

Corn stack	2–3
Cutting bonds	1*
Feeding	1
Clearing chaff and cavings	1
Straw rick	2
Sacking grain	1
Engine or tractor	1

* Very occasionally 2. Sometimes the feeder cuts the bonds.

A friendly word of advice to the novice is that he should avoid the task of clearing the chaff and cavings, which is very dirty and uncomfortable work; it usually falls to the lot of a boy who cannot foist it elsewhere. Cutting the bonds is a hard task, and in some cases the work of the whole gang is quickened considerably by allotting two men to the work; the bond of each sheaf should be grasped at the knot and a cut made close to that point so that the string comes away easily. In some districts it is usual for the feeder to cut the bonds, so that a man is economised, though this must tend to slow down the work of the whole gang. Rarely are sufficient men allowed for the straw rick, which is consequently badly built so that rain soaks in and spoils the straw. Of the two men who accompany threshing sets, one feeds the drum whilst the other supervises generally. In addition to the above gang, other men may be required for carting away the grain. With so many men involved it is important that no time should be lost; if possible the engine, thresher and elevator should be set in position overnight, whilst the placing of stacks in pairs, so that two of them can be threshed at one setting, economises much time. The order of threshing must be arranged in view of the fact that an hour or so may be occupied in clearing up, when changing from one crop to another.

The amount of corn threshed in a day varies between wide limits; factors affecting it are the standard and organisation of the labour, the length of the working day, the efficiency of the threshing machine, the length of straw and the size and fullness of the ears or pods. In British practice typical amounts (in bushels) threshed in an 8-hour day would be:

Wheat, beans and peas	350
Barley	400
Oats	500

These figures compare poorly with those of other countries, in which larger machines with greater width of drum are commonly used; the world's record is in excess of 6000 bushels and this would be impossible to achieve with the ordinary thresher. Peg-drums have been tried in this country but have not found much favour, because they produce straw which is very broken; the self-feed and the disposal of chaff by blowing have, however, been appreciated. It is possible to incorporate these features into British machines and then, with careful organisation, threshing speeds might be materially increased and, at the same time, the size of gang reduced.

WINNOWING (OR DRESSING)

Modern threshing machines deliver good samples of grain and, in general, no further cleaning is necessary before sale; with wheat, oats, beans and peas winnowing is rarely carried out except where the grain can be sold for seed at a high price. With barley, however, winnowing is fairly common, as it improves the evenness of the sample, which is of the greatest importance to the maltster. After threshing, the grain is accumulated loose in a heap on the barn floor, and this mixes the barley from different parts of the stack, and hence of the field which grew the crop; this is important as tending to even out local irregularities due to soil variation, the presence of trees and so on. It is very necessary for the producer of high-quality barley that he should gain and keep a good name with the dealers and maltsters in his neighbourhood; it is not enough that the small pocket sample on which the sale is made should be a fair one, because it is further important that each sack of the bulk which he delivers should be true to the sample. Winnowing goes far to ensuring this, besides raising the general level of quality by removing light and small grain. Winnowing is work very suitable for wet weather; three or four men are required, and they should deal with at least 200 bushels in a day. Often, however, the work is done during the dark days of winter, in a barn with no artificial, and very little natural, light. If it were worth while to provide good conditions for work, and to mechanise the operation, the daily out-turn could be increased immeasurably; but most farmers have no great amount of corn to winnow, and the interest on the money necessary for full equipment would probably exceed the annual financial saving.

CHAPTER X

HARVESTING BY COMBINE

The harvesting methods described in the last chapter are the traditional ones and, though the work has been mechanised stage by stage, the procedure has remained fundamentally unchanged for centuries. The weakness of those methods lies in the repeated handlings of the material; sheaves must pass through human hands at least six times before they come to lie in a stack and the straw is handled at least six more times before it is in a straw rick. Combine harvesters (in common parlance usually referred to as 'combines') revolutionise the whole procedure and their use avoids this repeated handling. In the original homes of combines, America and Australia, agriculture is characterised by dear labour and cheap land, and it was the higher cost of labour after the War of 1914–18, as compared to pre-war days, which led to combines being seriously considered for British farms, albeit that land in this country is not cheap. First introduced in 1928, combines have spread into all British counties and in the arable districts are now regarded as the normal means of harvesting grain. When advanced practitioners of the new methods debate, their talk runs on lines which seem weird and wonderful to older farmers, such matters as the automatic stoking of furnaces assuming great importance; the discussion here will not cover engineering details, important though they are, but will be limited to general principles.

As it travels round the field the combine both cuts the corn and threshes it, delivering grain and straw. The most impressive models are self-propelled with cutter bar in front, so that they can be driven straight into standing corn without crushing any down; controlled with easy precision by one man, they are very manoeuvrable even with a cutting width of 12 ft. For all farms except the very large ones a cutting width of 8 ft. or 8½ ft. is generally reckoned enough and such a width is convenient as the machine can be moved easily from field to field or along a public highway. On smaller farms cutting widths down to 5 ft.

or even less are usual. All large machines, with cutting widths of 8 ft. and upwards, have petrol engines built into them to move the working parts. Smaller machines may be driven partly by the power-take-off from the tractor that pulls them, and partly by gearing from their own travelling wheels. The grain can be delivered through a sacking apparatus, in which case full sacks (usually of only 2-bushel capacity) are left in groups of four or five over the field; alternatively, a tank may be fitted, in which the grain is accumulated and from which the loose grain is emptied into trailers or lorries at intervals. There is no doubt that handling the grain loose saves a great deal of labour; picking up sacks of grain in the field is slow and heavy work. Nevertheless there is something to be said for sacking the grain where the acreage to be combined is small. Loose grain cannot be accumulated beyond the limits of the intake hoppers at the drying plant (see below) and requires a steady stream right through to the final storage; even if it is rather damp, grain takes no harm in sacks for a day or two so that when it is sacked on the combine it is under control from then on. Normally the machine is used for 'straight' combining, that is, for cutting and threshing. If, however, the crop contains much green material the sieves of the thresher become repeatedly clogged and many stoppages occur; in that case it may prove desirable to cut the crop first leaving the cut material lying in a loose condition on a long stubble. After allowing the green plants a few days to wither, the combine, with a special attachment instead of a cutter bar, is taken along the wind-row, which it picks up and threshes; the pick-up attachment is wonderfully efficient, but the wind-rowing method is not common.

The way of the pioneer is hard and the early users of the combine in this country received a full measure of derision; it is true that they invited this, by trying to adapt British agriculture to an American machine, but the futility of such a reversal of common sense soon revealed itself. The old-fashioned farmer, who laughed heartily at the idea of using combines in this country, was not wholly unjustified in his merriment, in view of the many grave objections which his experience suggested; it is true that some of these points have proved less serious than he imagined, but others have been met by modifying methods associated with combines, and the combines themselves. It will be instructive to examine the objections at some length.

(*a*) *Cost.* At the present time combines range in price from about £600 for a 5 ft. cut up to nearly £1800 for a self-propelled 12 ft. cut machine (with one model costing as much as £2475). This is only part of the outlay required when a farmer turns over to combining (see p. 303) and initial cost must remain a formidable difficulty in this country of relatively small farms. The traditional method of meeting such a situation in British agriculture is the system of contractors; that is, moneyed men, often themselves farmers, purchase expensive machinery and let it out to farmers on a piece-work basis. The system has worked well with threshing sets, steam ploughs, etc., but would not appear very suitable for the combine whose season of usefulness is very restricted. A combine contractor leads a hectic life during the harvest because nearly all his clients require his services at the same time, and if the weather is unsettled their demands get very insistent; despite this, combining by contract is extending. When a combine is being used only on one farm it is possible to extend its working season somewhat by planting early and late varieties, and thus spreading the ripening period; to arrange this among a group of separate farmers is impossible. It appears, therefore, that a combine can only find a place on farms with sufficient acreage of corn to justify its initial cost; on a small area the charge per acre for interest and depreciation on the machine would be prohibitive. The rate of working generally claimed for machines with a cutting width of 12 ft. is 2 acres per hour, but this can only be achieved under good conditions; upwards of 20 acres have been cut in a day, but stoppages for mechanical or climatic reasons usually reduce the day's work well below this acreage. The aim should be to cut 30 acres in a season for every foot width of cutter bar, but this is difficult to achieve. Some farmers have, by using their machines for a variety of crops and consequently extending their season, covered well over 360 acres with a 12-ft. combine. Clearly only large farmers can do this amount and even for a 6-ft. cut nearly 200 acres are required to reach the standard; allowing for grassland, root and green crops, this means a farm approaching 400 acres. There are those, however, who argue that the standard is unnecessarily high and that the saving of labour is so great that 80 acres of corn will justify a small combine. It is not always necessary to buy a new combine and then the above arguments do not apply. The advantages of the self-propelled front cut models have induced a number of farmers with large acreages of grain to change over to them from tractor-drawn models, which they then sell for what they will fetch; this is often quite a nominal figure. Some serviceable machines with a number of years work in front of them have been sold for as little as £25. At this price they may be a very good investment for a farmer growing very few acres of corn, especially if he can arrange to cut for one or two neighbours on contract.

(*b*) *Climate.* This was thought to provide an insuperable objection to the combine in this country; in countries where blazing sunshine could be depended on during harvest the machine was obviously suitable, but the fickle weather of this country was believed to make the combine a poor gamble, with the certainty of failure in some seasons. It has been found, however, that corn which is dry enough to cut with a binder is dry enough to cut with a combine, though, of course, the latter requires it in a more mature condition than the former. This argument has, in fact, been completely reversed, the combine owner claiming that it is in wet harvests that his machine shows to the best advantage; he only requires to have his corn dry once, whereas where older methods are used the corn must be dry for cutting, and then it is left out in the stook to get wet again. It is very significant that, whereas in the early days of combines they were restricted to the drier districts in the south and east, in recent years their numbers have increased most in corn-growing districts farther north; Northumberland is a county where combine harvesting has made great strides and these machines are rapidly establishing themselves in parts of Scotland. It seems that the pioneer has turned this argument effectively in his favour, except that by his methods harvest is postponed for a week or a fortnight, with consequent greater risk of the breaking up of summer weather and of storm damage.

Nevertheless, the climate of this country has imposed one penalty on the combine user—that is, providing himself with means for drying the grain. The combine is capable of threshing a crop satisfactorily when the moisture content of the grain is as high as 30 per cent, but such grain must be immediately dried to a level of about 15 per cent; wet grain may sprout, heat or mould, and if stored for only a few days it becomes unsaleable.

The principle of grain drying is that hot air is blown through the mass of grain, and that when sufficient moisture has been removed cold air is blown through to cool the grain, as otherwise there is considerable condensation of moisture during cooling. An obvious danger is that the temperature may be raised too high, and that the grain may be killed; a thermometer inserted in the grain should never show a temperature higher than 140° F., and most farmers would prefer to keep to safer levels than that. There have been disastrous cases with barley, which has been sold for malting purposes and then found to have had its germinating power destroyed; with that crop it is now usually recommended that the temperature of the 'air plus grain' should never exceed 110° F. Nor should drying be continued after the necessary amount of water has been removed, because otherwise a cooked smell is imparted to the grain. Grain drying has passed very quickly through the trial-and-error phase and now, with quick methods of determining moisture content of grain and continuously recording thermometers at the dryer, can be said to be on a scientific footing. The safe maximum temperatures for various types of grain and seeds are now fairly well known. There is no reason why the quality of grain should in any way be impaired by drying, though some maltsters are still loath to buy barley that has been through a dryer. Several different makes of dryer are available, the type most favoured at present being that in which the grain passes slowly down narrow vertical columns, with hot air blown through the top two-thirds of the height and cold air through the bottom third. The furnace may be for coke, anthracite, oil or gas, whilst electric power is most convenient for the moving parts of the outfit. It is best to pass the grain first over a precleaner which sifts out things like thistle heads and thus saves the drying of green material which is very inflammable when dried; after drying, the grain goes to a dressing machine which produces a first-class sample. There should be two large hoppers to receive the grain as it comes in from the field, two rather than one so that grain from separate crops may be kept apart. Where the combine is fitted with a grain tank the grain is carted loose in lorries or trailers which can be tipped at the reception hoppers; this saves handling but as noted above gives less flexibility than where the grain is sacked off the combine.

Some farmers have made their own tray-dryers at small cost, but efficiency of drying is generally less than where a dryer of established reputation is installed. It is common to have conveyors so that movement from the reception hopper to the pre-cleaner, from there to the dryer and thence to the dresser and finally to storage is all automatic. A dryer may be purchased for less than £1000 but, with the cleaners and conveyors, by the time it is installed the cost is not likely to be much less than £2000. Cheaper forms of dryer are becoming available and further progress in this direction may make the installation of a dryer economic for farmers with moderate acreages of grain crops. Ventilated bins provide storage as well as drying; the initial cost for the drying equipment for these bins is around £500 for a plant of 100-ton capacity. In-sack dryers usually have lower initial cost than ventilated bins but this depends on the number of 'holes' they have (that is, the number of sacks that can be dried at once). Dual-purpose dryers that can deal with grass during the early part of the summer and grain at harvest time have obvious advantages, and with very little adaptation in-sack dryers can do both jobs, but as yet few are used in this dual capacity.

There are only a little over 5000 dryers on British farms and so there must be many farmers who rely on combines to harvest their grain but have no dryer of their own. In a fine, dry harvest all may go well with them but generally their combines cannot cut a large acreage in a season, because they have to await dry conditions and, in particular, cannot start cutting much before midday. Some firms and official bodies have large dryers and take undried grain from farmers, which greatly helps the extension of combining to smaller holdings. It is very risky to face a harvest with a combine and no arrangements for drying; if the risk is taken sacks may be put under cover with their mouths open for several days but if the grain has a high moisture content it soon deteriorates. A few farmers leave the sacks in the field, standing them up in the hope that there will be a drying wind. These devices are not very satisfactory and in a wet harvest a regrettable proportion of grain is delivered in bad condition.

(*c*) *Straw.* In mixed farming straw plays a large part, both in regard to stock and, when returned to the land as farmyard manure, in maintaining the fertility of the soil; the greatest saving can be effected by the combine when the straw is not required and, in the countries where the combine is well established, a common practice is to burn the straw. The larger combines are incapable of dealing with long straw, partly because of the narrow platform on which the cut material falls, and partly because in a heavy crop the threshing drum becomes blocked; these machines, therefore, leave a stubble at least 6 in. longer than that left by the binder, and naturally this greatly reduces the yield of straw. Another reason for cutting high with a combine is that the straw-walks on present models are barely large enough, in length and width, to ensure that all loose grain is shaken free; by reducing the amount of material that goes through the machine something is done to avoid the loss of this grain with the straw and, of course, less straw means greater speed of working. A spreader may be fitted to a combine to distribute the threshed straw, as it comes from the shakers, evenly over the ground and this is a convenience if the straw is to be ploughed in. More generally the straw is left loose in a wind-row and various methods have been used to collect it. Sweeps may be used, in conjunction with an elevator or stacker, to build a loose rick in the corner of the field, but baling methods are steadily gaining favour. It has not yet been found possible to trail a baler behind the combine to take the straw direct from the latter, largely because of the extra draught this involves. One obvious solution is to sweep the straw to a stationary baler in the middle of the field, but farming opinion is hardening in favour of the one-man pick-up baler. A pick-up baler, working 2 or 3 days behind the combine if there is much green material to dry out in the wind-row, can easily keep pace with a large combine, but considerable labour is necessary before the finished bales are neatly stacked. Combined straw need not necessarily be burnt or ploughed in, as was too commonly the procedure of the pioneers, but its collection involves expense which greatly reduces the difference in cost between combine and ordinary harvesting; that the value of the straw may be more than sufficient to cover its cost of collection does not detract from the argument.

(*d*) *Heavy and varied crops*. In the regions of the world where combines have proved successful, farming is extensive and yield per acre is very low compared to this country; it was felt that the machine could not handle the amount of material commonly grown in this country. Experience has shown, however, that heavy crops can be dealt with satisfactorily; a swathe narrower than the full width of the cutter bar can be taken, but in practice this is rarely found to be necessary.

A machine to be worth its place in a system of mixed farming must be capable of dealing with a variety of crops and it was thought that the combine would fail in this respect, restricting its owner to a few selected crops, which would lead to unsound agricultural practice. This was, in fact, one of the earliest effects of introducing combines but they have now been used, with some minor adjustments, for nearly all the crops harvested in this country for seed, and successfully for most of them. Wheat and barley are probably the most suitable crops for combining and of these the latter is slightly preferable, because it is normally cut when dead ripe and despite the fact that its awns may block up sieves. Rye does not present any difficulty, but the oat crop is not a good one to combine; oats cannot be threshed in the immature state in which they are normally cut by binder, and if they are left to grow until they are fully ripe the grain is liable to be blown out of the bells. The bean crop is one of the few with which the combine cannot deal satisfactorily, since the corn shatters badly when beans are dead ripe and until then they carry much green leaf. Peas, on the other hand, are frequently combined. A very good method is to cut them with a machine built round a tractor and having blades running just below the surface of the ground; torpedo-dividers run in front of the tractor wheels and the cut material is left in wind-rows which a combine, with pick-up in place of cutter bar, threshes a few days later. Root seed crops, sunflowers, grasses and clovers have all been combined satisfactorily; red clover seed has, in fact, provided a triumph for the combine and the old laborious methods, always attended by risk of utter failure in the short days of September and October, are likely to pass out altogether. The wide and surprising adaptability of the combine has been an important factor in its progress; it is no longer necessary to concentrate on one or two crops for it, the variety of crops extends its season and moderate-sized farms working to sound rotations can find enough acres for it.

(*e*) *Loss of corn.* Where barley is to be cut with a combine the cutting need not be delayed longer than the normal stage of ripeness for the binder, since that is the dead-ripe stage. As has just been mentioned, oats cannot be left much longer than usual because of the danger of loss of grain. With wheat, the general advice is that the crop should be left about 10 days longer before cutting with a combine than with a binder; the grain must be flinty hard before the crop can be threshed cleanly. To leave a crop standing longer than normal exposes it to weather risks and some loss of corn must be expected, but it must be remembered that where harvesting is by established methods considerable shelling often occurs during carting. Careful tests have shown that losses of shed grain often amount to 4 or 5 bushels per acre, but this frequently occurs when older methods are used; clearly there must be great variation from case to case, but the available evidence indicates that losses are in general rather less where harvesting is by combine than where it is by binder.

(*f*) *Size of fields.* An unwieldy implement like the combine requires a certain amount of room to manoeuvre, but a corner can be turned with it practically as quickly as with any other machine. The initial expense of the combine requires that it shall be used economically, so that it is more important to avoid waste time with it than with cheaper machines. Most models can, in a short time, be reduced in width sufficiently to pass through a gateway 11 ft. wide; unfortunately many of our gateways have a width of only 9 ft. A self-propelled 8-ft. cut machine can negotiate, unaltered, all ordinary gateways and is more handy than a binder; this width of cut has become very popular and such models can operate economically in a district of small fields and stockproof hedges. Machines with wider cuts are admittedly more suited to open land where hedges are of little concern, but the grubbing of a few hedges will not be a deterrent if their presence militates seriously against the spread of combines. Steep slopes are usually cut up and down rather than across, but in other countries hillside models of combines are used which allow the cutter bar to conform to slopes whilst the thresher remains level.

(*g*) *Marketing the grain.* The combine farmer has all his grain ready for marketing immediately after he has finished harvest, whereas with normal methods a proportion may be held in the stack for months. At present the guaranteed price for wheat rises, roughly at two-monthly intervals, from £27. 10*s.* per ton at harvest time to £32. 10*s.* in May and June. Prices for the other cereals are not fixed in the same way and seasonal trends are very variable but in general the tendency is for them to rise through the winter and spring. A farmer unable to hold his grain is in an unenviable position since he cannot take advantage of the assured rise in wheat price nor use his judgement as to when to market the other cereals; this consideration is inducing ever more farmers to erect their own storage. Cylindrical silos, up to 15 ft. high, built of wood, concrete, brick or steel, are the usual form of storage and it is common to have a battery of six or more silos. Some conveyor system is necessary so that grain may be moved as required from one silo to another, in order to aerate it and prevent the development of heat or mould; the grain may be conveyed pneumatically or mechanically. If all hand-work is to be avoided the silos must be funnel-shaped with the inside walls sloping at an angle of at least 45° to the opening over the conveyor; this wastes much space and sometimes it is preferred to have a flat bottom and shovel out the last few hundredweights of grain when a silo is emptied. Storage and conveyor systems entail further initial expense for a farmer turning over from the binder to the combine; this outlay is usually of the order of £10 per ton of storage space, and it is desirable to have sufficient to hold at least half the grain combined at a harvest.

It will be appreciated that some courage and much capital is required before a farmer abandons the traditional methods of harvesting corn, for which he will be fully equipped, and turns over to combining. The outlay involved will vary greatly from farm to farm, since some have dryer, pre-cleaner, dresser and storage with enough conveyors to make every movement of grain entirely automatic, whereas others purchase little new equipment apart from the combine. The full equipment for a farm expecting to combine 400 acres of corn in a season would cost somewhat as follows:

	£
8½-ft. self-propelled combine	1600
Dryer, cleaners and installation	1800
Storage silos and conveyors	2600
	£6000

This is a very formidable sum and it is by no means complete. It would scarcely provide storage for all the grain from 400 acres of corn and probably the total expenditure would be seriously increased in adapting an existing barn or, worse still, in erecting a new building to house the dryer and storage. A pick-up baler would be required but the farmer may already possess one for haymaking.

Thus far the argument for the combine has been presented in a negative manner; it has been urged that the objections which practical men in this country raised against it have not proved so formidable as was expected. The positive part of the case for the combine lies in the question of cost and in the saving of labour. By the older method the cost of cutting, stooking, carting, thatching and threshing normally amounts to £7 to £8 per acre; where the corn is threshed from the stook, eliminating stacking and thatching, the figure is reduced by about £2 per acre. It is very difficult to arrive at average cost for combining. This is not because actual costs are unobtainable, for current agricultural literature abounds in them, but because conditions vary so much from case to case and because price levels have been continuously rising in the period when most experience has been gained. One very uncertain factor is the proportion of grain that must be dried; in a favourable season this may be less than 20 per cent, whilst in a wet season it may exceed 80 per cent. When the variation in equipment and in acreage cut, and hence the interest and depreciation charge per acre, is remembered, it is clear that costs from actual cases may not be very helpful, and hence it is proposed only to make general statements; no figures can have absolute value but it is thought that the following are valid for a rough comparison with the cost of harvesting by binder. Field costs for combining, including the haulage of the grain to dryer or barn, amount to some £3 an acre whilst drying and dressing may add £1. 10*s*. Where straw is burnt or ploughed in, therefore, combining shows a saving of about £1 an acre as compared to cutting by binder and threshing direct from the field with another £2 an acre to be added to the margin where the corn is cut by binder and stacked. For a fair comparison the expense of collecting the straw should be added to the costs for combining, an addition of at least £1 an acre; it could be argued that this is an optional cost with combining but obligatory where the binder is used. But that is not the whole story. In the expenses for combining, interest and depreciation on the costly equipment loom very largely, so that the greater the acreage of grain crops a farmer has the more are the scales tilted in favour of combining. With traditional methods labour accounts for most of the cost. In round figures, combine harvesting calls for some 100 man-hours of labour per 10 acres, which compares with 150 for binder and direct threshing and nearly 250 for binder and stacking. For the last 10 years labour has been scarce and these differences have weighed quite as heavily with farmers as differences in working costs.

The establishment of the combine in this country has been rapid. In 1928 there were only two operating (including one imported solely as an experiment) and from then till 1939 was a period of trial and some errors, in which numbers increased slowly so that by 1939 approximately 100 machines operated. During the War of 1939–45 the corn acreage was greatly extended and labour was scarce and dear so that it is not surprising that the rate of increase was quickened; it was not only that enough hands to harvest the war-time crops were difficult to find but also that with the older methods there seemed endless threshing to be done during the short days of winter. In the harvest of 1948 over 6000 combines worked in Great Britain and in recent years the spread of combining has accelerated so that some 30,000 combines worked in the harvest of 1956. For that harvest it was estimated that well over half the acreage of wheat and of barley was combined but only about one-sixth of the acreage of oats.

In view of the high yield per acre it is to be hoped that corn-growing will always find an honoured place in British farming and if that is so there can be no doubt that the combine will continue to extend its sway. The initial expense in adopting combining must always be high, but once a British farmer is equipped he should be able to produce grain as cheaply as any farmer in the world. There remains for solution the question of straw. It has become common to refer to this problem as one of straw disposal—as though straw were a necessary evil and the farmer's problem how best to rid himself of it. Straw can be ploughed in to maintain fertility of the soil and this is infinitely preferable to burning it; machines to chaff straw from the combine, methods of undersowing corn crops with trefoil, or of sowing stubbles with mustard, to get some green plant to grow through the straw and anchor it, so that the plough may cover it more easily—all these have been tried and more will probably be heard of such matters. But the ideal solution cannot be found in these directions. Straw is not a nuisance, but a valuable by-product of corn-growing and one of proved worth in livestock husbandry. The problem is, therefore, to devise cheap and efficient methods of collecting the straw left by a combine, and present indications are that a pick-up baler producing string-tied bales and operated by one man provides the most practical answer.

In the first flush of enthusiasm the combine pioneers cast aside all the knowledge of the fundamentals of British farming so painfully acquired by their ancestors. Mechanisation was their watchword, and it became almost a point of honour not to have an animal on their farms. Impressed by the appetites of their machines, they speculated on cereal growing and tended to ignore the dangers, of impoverishing their land and of disease, which were inherent in their systems; it is scarcely a matter for wonder that older farmers spoke with withering scorn of farming for cereals, in some cases for wheat, alone. But that phase rapidly passed and it was soon seen that combine farming could not survive if it meant what by all canons was bad farming. The combine has made more healthy and assured progress since it has been adapted and incorporated into what is still the mainstay of British agriculture—mixed husbandry.

CHAPTER XI

THE HARVESTING OF ROOTS AND POTATOES

Whereas the corn harvest is a definite entity, concentrated into a short period and capable of description in a general manner, the harvesting of roots and potatoes occupies very varying lengths of time during the autumn and early winter months, and involves diverse methods for the different crops. It will be necessary, therefore, to divide this chapter into sections and to deal with sugar beet, mangolds, other fodder roots, and potatoes separately.

SUGAR BEET

The harvesting of sugar beet is long-drawn-out and very costly. If the farmer were a free agent in the matter, and could lift his sugar beet when he chose, the following factors would determine for him the optimum time:

(*a*) *Yield of sugar*. As the autumn progresses sugar beet gain in weight of root, and in sugar percentage of root, until the end of October or early November; throughout November little change occurs, but during December there is normally a serious fall in the percentage of sugar of the root. Weight of root rises, on the average, by 10–20 per cent from late September to November and then remains constant; sugar content rises by as much as 2 per cent (e.g. from 15 to 17 per cent) in the same period, but may decline by a like amount, or even more, during December. The above is the general rule but individual crops vary much. The weather of the season affects the rise and fall to a great extent; for instance, after a very dry summer weight of root may be nearly doubled during October, this being attended by a considerable, though mainly temporary, fall in sugar content. Soil rich in humus, heavy nitrogenous manuring, late sowing and late maturing varieties are all factors conducing to a late optimum sugar yield.

(*b*) *The crop which is to follow the beet.* If it is proposed to grow wheat on the field in the following year it is desirable that the sugar beet should be cleared early; wheat often does follow sugar beet and there is ample time in the case of the first cleared field of a large grower of sugar beet. It is more common, however, for a spring-sown crop, usually barley, to follow beet and then this consideration does not arise.

(*c*) *Difficulty of carting late in autumn.* It is to be expected that as the depth of winter approaches the land will become wetter and softer; thus late lifting means heavier labour in carting, greater damage to soil texture, and higher dirt tares.

(*d*) *The use of the tops.* Sugar-beet tops are valuable food, and it may be desirable to start lifting in good time to provide for the sheep flock or for cattle.

The main consideration is, of course, yield of sugar, and in order to obtain the optimum in this respect the farmer would prefer to lift as much of his crop as possible about the first fortnight in November. But a sugar-beet factory must keep running for a considerable time each year if the capital invested in it is to earn reasonable interest, and, furthermore, if all farmers delivered a large proportion of their beet at about the optimum time the factories would be submerged. This is averted by a clause in the contract between the factory and the farmer which is signed before a sugar-beet crop is sown. This clause is that beet must be delivered in about equal quantities per week from October 1st to December 31st. In actual fact the campaign opens at a date, determined by the factory, from about September 20th to October 10th, varying with the season and the estimated throughput that the factory has to face; the campaign finishes when the crop runs out, this date varying from about December 10th to the end of January. When a factory opens, its growers are notified and 'free loading' (that is, farmers may send in their beet as fast as they like) is allowed until large reserves have accumulated in the factory's flumes; after that each farmer is allotted weekly permits, the number of which is determined by the acreage of beet for which he has contracted. When the factory's field men see that the crop is nearly all delivered a closing date is fixed, and growers are again allowed free loading, so that all their crop may be at the factory in time. Thus the farmer can deliver beet as fast as he chooses for a short period just after the factory's campaign opens and again, if he has any left, just before it closes; during the majority of the time, however, his weekly deliveries are restricted according to the acreage he grows. Naturally there is considerable flexibility, factories allowing extra permits where a farmer's crops are very heavy, and in special circumstances; such cases are arranged amicably between the farmer and the factory manager, but in general the latter insists on the appropriate weekly limit, as he must if he is to avoid being 'snowed under'.

The only choice open to a farmer then, is whether or not he will take full advantage of the free loading at the opening of the campaign, and of his permits week by week; if his crop is light he may postpone starting till well into October, but in happier circumstances he must begin early and keep up with his permits, if he is to avoid much delivery at the end of the campaign, when weather will probably be inclement and sugar content will have declined. It might appear that temporary clamps would be worth while, so that the beet could all be lifted near the optimum time for sugar yield, and then delivered as permits allowed; if this plan were adopted widely the factory could be worked for a great part of the winter with consequent rise in efficiency. But clamping of beet leads to large losses. The beet continue to respire in a clamp, and this means a steady and quite appreciable loss of weight as they remain there; furthermore, moulds and decay reduce the weight to a large extent, particularly in some seasons. Thus clamping is to be avoided if possible, except for the very temporary accumulation of beet at a convenient point for loading into lorries, etc.

The operations involved in the harvesting of sugar beet are lifting (that is, loosening the beet but leaving them in the ground), pulling and knocking, topping and delivery. Good progress has been made during the last twenty years in the mechanisation of these operations and reference will be made later to beet harvesters. It is necessary to know, however, what is involved in the beet harvest so that it will be convenient to deal first with the various stages of the work when carried through by hand labour.

LIFTING

On very loose soil, and when the crop is a poor one with its roots penetrating no great distance into the ground, it may be possible to pull the beet up by hand, without any preliminary loosening; in general, however, this involves very strenuous labour and much blasphemous kicking, because the leaves are often pulled off and the root left unmoved. On small areas lifting is sometimes done with a fork or spud, which is pressed into the ground, the beet being moved by a levering action; the work is not very hard nor unduly slow, but sometimes the roots break laterally and an appreciable proportion is left in the ground. A common plough may be used to turn a small furrow away from one side of each row of beet, which can then be pulled sideways and up fairly easily; this method is unsatisfactory, however, and is rarely adopted. Some big growers have used the heavy cultivator for lifting beet, but this practice is now very rarely encountered; the bursting action of the tines loosens the roots reasonably well, but subsequent carting is rendered heavier since the soil is broken up to a considerable depth. Usually a sugar-beet lifting plough is employed, and there are two types in common use. In one the plough body consists of a deep plate carrying at its lower end a share, sloping upward and backward; the share runs under a row of beet and its incline forces each beet upwards an inch or so. In good conditions this implement does extremely efficient work, each plant being moved slightly and rendered quite loose and easy to pull; it is important, however, that the share should not be badly worn, because then many beet are left unmoved. If the land is very wet the plants tend to be pushed sideways and smeared with mud, but even then subsequent pulling is easy. On the other hand, if the land is very dry, and particularly if it contains an appreciable amount of clay, it is very difficult to get the plough into the ground and to keep it in; under extreme conditions the operation is practically impossible with any implement, and the only thing to do is to wait until rain comes. The other type of lifting plough consists of pairs of plates, one of each pair running on each side of a row of beet, which squeeze the beet upwards. This squeezing action tends to paste mud on to the roots which are thereby rendered more difficult to clean. On light, loose soils the pasting action is not serious and the single share type tends to run away from its work; this type has certainly done good work in the past, when horses were commonly used, but is now becoming rare. The double-wedge type is used exclusively in beet harvesters since it will force the beet right out of the ground, whereas the single inclined share does no more than loosen the roots.

With either type of lifting plough two horses are sufficient, and 2 acres should be covered in a day's work; where a tractor is used two or more double-wedge type of lifters are fitted to the tool bar. Lifting is the most rapid part of the harvesting operations and consequently it is not necessary to continue it regularly throughout the campaign; on ordinary-sized farms odd days, or half-days, are sufficient to keep the lifting ahead of the pulling. It has been asserted that considerable gain accrues from keeping an interval of about a fortnight between lifting and pulling, because maturation is hastened and hence sugar content raised; surprisingly little experimental work has been done on the point in this country, but such results as are available fail to indicate that anything is to be gained in this way.

PULLING AND KNOCKING

When pulling and knocking are done by hand, the method is to take two rows of beet at once, to pull a beet from each row at the same time, knock them hard together once or twice and lay them down, with all the tops in one direction in a single row; by laying them to one side and putting the beet from the next two rows on top of them, a row of knocked beet is formed for each four of the original crop rows. Efficient knocking is very important because otherwise dirt tares will be high.

TOPPING

Pulling, knocking and topping are very commonly done by piece-work, the men pleasing themselves as to how much of the first two they do before performing the last. When topping, a man works along one of the composite rows of knocked beet, throwing the topped roots to one side to a row of heaps; if the same row of heaps is used in working back across the field, it will eventually contain the topped roots from eight of the original crop rows. The most common method of topping is to use a light chopper and to stand straddling the row; a plant is picked up by the point of the chopper, held in the left hand, topped with one stroke of the blade and thrown into a heap. Where the roots are very large many fall off the point of the blade as they are picked up, and also it is difficult to hold one in the left hand and to cut right through the crown at one stroke; in that case, therefore, it is common to top them as they lie on the ground, the same chopper, or perhaps a sharp spade, being used. A clause of the contract made with the factory stipulates that the tops shall be cut off just below the point where the lowest leaf grows from the root; thus the severed tops consist of the green leaves and the crowns of the roots. It is important that topping should be done accurately. If the root is topped too high (and if it happens to be one taken in the sample at the factory—see below) the piece above the lowest leaf is removed and becomes part of the tare. At the factory the crown is not required because of its low sugar content, and the impurities it contains which render difficult the crystallisation of the sugar; from the farmer's point of view leaving too much crown on the root means having higher tare, extra cartage without extra return and loss of feeding and manurial value. On the other hand, if beet are topped too low the weight of crop delivered is correspondingly reduced. With practice men become very quick and accurate in topping beet, but it behoves the farmer to give careful supervision to the work.

The total cost of lifting, pulling, knocking and topping by the methods described above is around £10 an acre, most of it for man labour; on many farms extra, casual labour is hard to find for this work which is strenuous and occurs when weather is often inclement, and if it is done by a farmer's own staff some other operations on the farm must suffer. So it is not surprising that great efforts have been made to produce machines which will do the whole job. Very real difficulties have confronted the engineers. What is required is a machine which will, on all sorts of soil and in all weather conditions, lift the beet, top them accurately, clean them without damaging them and deliver the tops in a clean state for feeding. It cannot be claimed that all the requirements have been fully met but there are now a number of machines which approach the ideal. Progress in the last ten years has been very rapid. In 1946 less than 1 per cent of the beet crop was harvested mechanically but by 1956 this percentage had risen to over 50. It would be difficult to say how few acres of beet would just suffice to make it worth a farmer's while to get one of these machines and anyway the information would not be very useful. The long seasonal spread of the beet harvest favours contract work and a farmer with few acres of his own usually has little trouble in arranging for his machine to be used on the crops of some of his neighbours. Machines work much better in dry than in wet conditions so that mechanical harvesting favours earlier lifting even if that means some sacrifice of yield and temporary clamping before delivery. Some loss of yield is normally suffered in mechanical harvesting. Missed roots may amount to a ton weight per acre and, though machines adjust themselves to the height of each beet, where the plant is irregular the adjustment is far from perfect so that some roots are topped too high and some too low; this may be overcome by the principle adopted in a few machines of lifting the beet first and topping them as they are carried upwards suspended from their leaves. Some machines run over the tops and leave them pasted with mud and of little use for feeding but these and other imperfections will undoubtedly be corrected in the future and it can be confidently predicted that within a few years the great majority of beet crops will be harvested mechanically without any serious loss of yield.

DELIVERY

Moving the roots, from the heaps to which they are thrown after topping, to the factory is a very expensive matter. A few farmers are situated close enough to the factory to permit loading on to trailers which are taken straight to the factory; the vast majority, however, have to deliver by rail, by motor-lorry or by barge. In the case of rail delivery the farmer first has to cart the roots to the nearest station or siding by lorry or by tractor and trailer. The displacement of horses by tractors has certainly speeded up this road work but for pulling the beet off the field there is still much to be said for horses; tractors often get bogged down and where a farmer still has a horse or two they can well be used for this work. Some mechanical harvesters deliver the beet into a trailer travelling alongside them, trailers being drawn off the field as they are filled to a temporary clamp beside a hard roadway. The rail fare is paid by the factory but, of course, the amount is deducted from the cheques received by the farmer for the crop. The actual freight charges for rail delivery compare favourably with those for other methods of delivery but the intermittent work of carting from field to station as trucks become available is often a serious hindrance to other urgent work on the farm. The rail charge per mile is graduated steeply downwards as distance from station to factory increases; the following illustrates the charges prevailing at present:

Distance from station to factory (miles)	Rail fare per ton s.	d.
10	8	4
20	12	2
30	14	4
40	16	9

The British Sugar Corporation pays any rail delivery cost over 40 miles in England and Wales. In Scotland there is a different arrangement. There the price paid for the beet is 1*s.* 9*d.* less per ton of clean beet than that paid in England and Wales and the Corporation pays (whatever the actual method of delivery) the rail fare from the farmer's nearest station; this averages 17*s.* 8*d.* per ton.

Some farmers own motor-lorries, and in districts where sugar beet is grown extensively there are usually a number of lorry owners who undertake the work of delivering beet to the factory; these are often quite small men who contract during the autumn for more work than they can reasonably perform, so that a serious drawback to delivery by road is that often the lorry cannot be obtained when required, and deliveries fall behind schedule. Unless the ground is very dry and firm, lorries cannot be loaded on the field, and the usual thing is for the farmer to cart his roots with his own horses or tractor to a dump by the side of the road, from which the lorry is filled. This has the advantage that moving twice, with an interval for drying and washing by rain, leads to a reduction in dirt tare. Prices are by no means standardised and often form a matter for bargaining between the farmer and lorry owner; often the latter can arrange to carry a load (e.g. beet pulp) on the return journey, in which case he can offer the farmer better terms. A reduction of about 1*s.* per ton is usual where the lorry driver is assisted by the farmer's men in loading from the heaps at the side of the road. The following are approximate costs per ton for road distances up to 40 miles, beyond which delivery by lorry is extremely rare:

Distance (miles)	Charge per ton *s.*	*d.*
5	9	0
10	11	3
20	15	0
30	17	6
40	20	6

It must be remembered that the distance covered by the farmer's teams is usually very materially less in the case of road than of rail delivery; the difference in this initial cost may very easily outweigh the difference between rail and lorry charges.

Sugar-beet factories require very large supplies of water and so they are built on, or near to, banks of rivers; a farmer whose land adjoins the same river as the factory for which he contracts can deliver by barges which are owned by the factory. This used to be a cheap and efficient method of delivery but nowadays only about 1 per cent of the beet crop is sent in this way; the main method of delivery is by road (81 per cent of the crop), railways carrying the remaining 18 per cent.

There is extremely wide variation between farms in the cost of delivering beet to the factory. A man who has to cart his beet 6 or 7 miles to a station, and then pay for a 30 or 40 miles rail journey, may find his total cost over £2 per ton: another growing beet 5 miles from the factory may only suffer a cost of about 10*s.* per ton. Thus if the yield of dirty beet is 13 or 14 tons, the difference may be over £20 per acre, much more than enough to change a profit to a loss or vice versa.

Sugar beet suffer no harm from moderate frost but if a very cold spell occurs towards the end of the campaign farmers are urged by the factory to give their beet some protection; if the roots have been very severely frosted the factory may even refuse to accept them. Little field heaps may be covered with tops but large roadside heaps should not be covered with straw, since some of the straw is bound to find its way to the factory with the roots, and there it will clog the works. The best procedure is to make the heap a large one and to lay the outside roots carefully to form walls in which the decapitated ends of the beet are outwards.

TARE

When a consignment of beet arrives at a factory a sample, not exceeding 56 lb. and not less than 28 lb., is taken and weighed; it is then washed by a scrubbing machine and each beet examined to see if parts of the crown which should have been removed in topping remain on it, any part discovered being cut off. The sample is then weighed again and the difference between the initial and final weights gives the amount of tare which is expressed in the form of lb. per cwt.; a proportionate deduction is made from the weight of the whole consignment, to give the weight of washed beet for which the farmer is paid (according to the percentage sugar, as determined by analysis of ten representative beet from the washed sample). It will be realised that a high tare means that the farmer has paid for the carriage of much material for which he will receive nothing, and that the part of tare due to faulty topping represents a further loss of feeding and manurial value. It is not worth while to wash the roots on the farm before delivery (and it is useless to wash a few and put them on the top of the load in the vain hope that they may be taken in the sample), but every care should be taken to keep tares as low as possible, by thorough knocking and efficient topping. In a dry early part of the season tares should not exceed 6 or 7 lb. per cwt., but as the land gets wetter, particularly if the soil is heavy, higher tares cannot be avoided, and 15 lb. per cwt. must then be accepted as satisfactory. Where the crop is harvested mechanically this would be a very good figure. In a very dry time the amount of dirt adhering to the roots should be negligible, whilst the roots will absorb water on being washed; thus it happens very occasionally that the tare is negative, and something is actually added to the weight of beet delivered. At the other extreme, when very diminutive beet are delivered from heavy land in a wet period, tares soar to dizzy heights; cases of 80 lb. per cwt. are not unknown, though it is difficult to understand the optimism of the farmer who expects to make a profit by delivering a consignment of which so high a proportion consists of the land he rents.

TOPS

There is no doubt that the most efficient method of utilising beet tops (i.e. leaves and crowns) is to fold them off with sheep; in that case they are allowed to lie as they fall when cut off until the sheep reach them. Fresh tops contain appreciable amounts of oxalic acid, and are unsafe to feed unless chalk be given with them; when the tops have wilted for 10 days or a fortnight, however, they are a perfectly safe feed. From the weight and composition of tops it appears that, taking crops of corresponding excellence, an acre of them is equivalent to half an acre of a fodder-root crop; in practice beet tops are probably worth rather less because, owing to the fact that they are lying loose on the surface of the ground, many are trodden in and wasted. When it is intended to cart tops off they should be accumulated in heaps as they are cut off and it is probably better to cart them before the roots; it is important to avoid taking dirt with them. Tops may be fed direct to livestock (after wilting or with the addition of chalk) or made into silage, care being taken in the latter case to fill the silo only a few feet at a time and to allow an interval for each layer to heat up before another layer is added. Tops will keep in good condition for some weeks in a long clamp of triangular cross-section, no covering being required. If they are not wanted for feeding—many beet growers have no sheep and few cattle—they are ploughed in, in which case they should be spread evenly over the ground first; their manurial value is considerable and amounts, judged by the artificial manure required to produce an equivalent effect, to £2 to £3 per acre. It must be remembered, though, that when folded by sheep they also produce valuable manurial residues; in fact the manurial value after sheeping is very little less than when they are ploughed in, unless the winter is a very wet one when the more available sheep residues are liable to be leached. In other countries sugar-beet tops are dried, and valuable stock food produced; this has been tried in an experimental manner in this country, but there appears no future for the process unless price conditions change very markedly. For the 1955 season it was reported that the tops were dealt with as follows:

Ploughed in	37·3 per cent
Carted off and fed	31·8 ,, ,,
Fed on the field	30·5 ,,

The small residual (0·4 per cent) was made into silage or (an infinitesimal proportion) dried artificially. Making silage from sugar-beet tops is a chancy business and farmers can hardly be blamed for not using this method of conservation. On the other hand it is distressing to note the high proportion for ploughing in as this is a waste of valuable food for livestock; as mechanical harvesting extends its sway it is to be feared that this proportion will become even greater.

When some strains of sugar beet are sown early a serious proportion of the roots bolt, that is, send up flowering stems; in most seasons there are a few bolters among early-sown beet, but if they only amount to 1 or 2 per cent of the crop little harm is suffered. It was quite common to see care taken to exclude bolters from the beet sent to the factory, but now they are generally topped and included with the rest. The roots of bolters have a lower sugar content than those of normal beet, though the difference is not so large as is generally supposed, and the risk of including them is that they may be taken among the ten beet on which sugar analysis is performed; if not, their inclusion raises weight without lowering the sugar percentage on which the farmer is paid, and the gamble is generally considered worth taking.

MANGOLDS

The harvesting of mangolds involves some important differences from that of sugar beet, as the following considerations show:

(1) There being no factory to think of, the farmer can lift mangolds when he chooses.

(2) Mangolds are slow maturing and increase in weight and dry-matter content to a greater extent during the autumn, and probably until a later date, than sugar beet.

(3) Mangolds are liable to 'bleed'—that is, the juice oozes out if the skin be broken—so that the roots must not be cut; this is further important because mangolds are usually stored for some time and moulds may appear on cuts. Keeping quality is also lowered by bruises, so that mangolds should be treated as gently as possible.

(4) Mangolds, except for the Long Red variety, grow on the ground rather than in it, and consequently they can easily be pulled up without preliminary loosening. The fact that the mangold root is much bigger than that of the sugar beet, and consequently that fewer are grown per acre, is another important factor making the harvesting of mangolds relatively cheap.

(5) Mangolds suffer more harm from frost than sugar beet, and should never be touched when frost is on them; they will not be greatly harmed by an ordinary frost whilst still growing in the ground, but heaps of pulled roots should be covered by leaves.

(6) In general, mangolds are grown on heavier land than sugar beet, so that the increased labour and harm to soil texture of carting, when harvesting is delayed into the late autumn, is more serious.

(7) As mangolds are very rarely sold they usually have to give way to cash crops, of which sugar beet is their great competitor.

(8) Being grown on rather heavier land mangold crops are more often followed by wheat than are sugar-beet crops.

The facts that mangolds are not hardy to a really severe frost and that, owing to their late maturation, they are unsafe for feeding till after Christmas, preclude the folding of this crop; it is invariably lifted and clamped for use during late winter and spring, or even in the following summer or autumn.

It is generally believed that mangolds may be lifted safely any time after October 10th and, after a dry summer, growth sometimes appears to be entirely completed by that date; but when lifted early they rarely keep well, and it is probably better to leave them in the ground for another month or so, even though that means running the risk of severe frost. In the southern part of the country November is the month when mangolds are generally harvested, most farmers being content if the roots are in a clamp by the end of that month; in the more northerly parts of the country farmers prefer to complete the operation by the end of October or early in November.

LIFTING

When Long Red mangolds are grown on heavy land it is sometimes necessary to loosen them with a fork or with a plough before they can be pulled up; but this is not a common variety, and in the vast majority of cases the root comes easily from the soil when a pull is given to the leaves. Some farmers insist that no knife shall be used for topping because of the importance of not cutting the root; the leaves can be easily twisted off by hand, though a long day's work produces sore hands. Generally a light chopper is used, but the leaves are cut an inch or two above the crown of the root; the method is to hold the chopper in one hand, to seize the leaves of a plant with the other, pull the plant out of the ground, and swing the leaves against the blade of the chopper in such a manner that the root is released to describe a neat parabola to a heap or into a cart or trailer. Some practice is necessary before the art is mastered. There are two methods of organising the work. One is to pull and to top the mangolds first, and to leave the roots in rows of heaps for a week or more before carting; the other is to cart direct, roots being thrown into carts as they are topped. Of the two methods the latter is the cheaper, since it involves one less handling of the roots, but it has several disadvantages. If the roots are left for several days in the little field heaps they lose water, and there is no doubt that their keeping quality is thereby improved; in addition they are bruised less by this method than when they are bounced into a cart as they are topped. Another important point militating against direct carting is that the horses or tractors are not so fully employed as when carts are filled from heaps of topped roots; in many cases a farmer has much for his horses or tractors to do during the autumn, and consequently he does not want to spare them for any longer than he need for carting mangolds. It is often much more convenient to send men for odd days to lift and top mangolds, and then to organise a team for a brief spell of carting, in which the horses or tractor will be fully employed. It must be insisted once more that the roots should not be cut, nor bruised any more than can be avoided; if forks are used for loading mangolds they should have nobs on the ends of their tines, and any natural inclination to scrape off mud, or cut off fangs, with a knife, must be curbed. The tops of the mangolds consist only of leaves, the weight of which per acre is very small; after wilting they are a good feed and may be folded to sheep, but generally they are ploughed into the ground.

CLAMPING

The site for a mangold clamp should be chosen with some care. A dry spot alongside a hard road is desirable, because there will generally be much carting from it during wet weather. Sometimes the clamp is made in the field where the crop was grown, but it is usually better to choose a site close to where the roots will be fed; very commonly the immediate vicinity of the cowshed is the most convenient, but the neighbourhood of a grass field may be indicated if the roots are intended for ewes just after lambing. The usual form of clamp is triangular in section, the width of the base being 9 or 10 ft. and the vertical height 5 or 5½ ft.; such a clamp will contain approximately a ton of roots per yard of its length. Whilst carting is proceeding one man is kept busy at the clamp. As each cart arrives it is backed up to the end of the clamp and tipped; the man clambers into the cart while it is in this position and forks out the roots. The two sloping sides are carefully built, many of the roots being laid one by one. As time allows, the man at the clamp covers the latter with straw, a layer of about 6 in. in thickness being given; this straw should be laid carefully as a rough form of thatch so that water may be shed, and it is a convenience if loosely tied battens of straw are available for the purpose. A few clods are put on the straw at intervals to hold it in place, but earthing up should be postponed for about a fortnight because respiration proceeds actively for that time, and unless there is free ventilation keeping quality will be lowered. Eventually a layer of earth, about 6 in. thick, is put over the whole clamp, except for the apex of the triangle on which the straw is left uncovered to provide ventilation; in cold districts the apex is also covered with earth except at intervals of a yard or two, where wisps of straw are left protruding from the earth. Covering a clamp with earth is slow and laborious work and one man will rarely do more than about 15 yd. of the clamp in a day; sometimes six or seven furrows are ploughed round the clamp to provide loose earth, but this does not accelerate the subsequent work with spade and shovel to any large extent. In spring the mangolds start growth, which entails a wastage of food nutrients; growth can be checked very considerably by removing the earth, but not the straw, from the clamp.

If mangolds are carefully handled, not cut nor bruised nor moved during a frost, they keep very well; because of this, and of the fact that they are large and a heap of them contains large air spaces favouring free ventilation, mangolds are often stored in great heaps with no appreciable loss. Some homesteads are provided with a mangold 'cellar' (not, in fact, always below ground) which is merely a large room that is filled with mangolds; it is in close proximity to the cows' winter quarters and consequently the work involved in feeding the roots is greatly lightened. Mangolds take little harm from frost in such a cellar, in which they are usually given no covering. Another fairly common procedure is to build a more or less ungainly heap against one end of a straw rick, and to throw loose straw over the heap; sometimes square heaps are built in the field with only a covering of loose straw and no protection for the sides, and even then the loss through decay is not very great, unless a really severe spell of weather occurs.

For a few years after the War of 1939–45 fodder beet enjoyed a spell of popularity, especially for feeding to fattening pigs; they were used to economise meals and with the end of feeding stuffs rationing the fodder beet acreage rapidly dwindled. Various strains of fodder beet had dry matter contents ranging from about 15 per cent (i.e. little more than some strains of ordinary mangolds) to over 20 per cent; those with the higher dry matter contents were deeply rooted in the ground so that they had to be loosened before they could be pulled. Fodder beet are very resistant to frost and they can be left in the ground until they are required, nor do they suffer from bleeding if they are cut.

OTHER FODDER ROOTS

Green crops are not usually harvested but are folded off on the field. Marrow-stem kale, however, is not infrequently used for green soiling to cattle in yards or at grass; for this purpose it is better if the kale has been singled so that few large, rather than many small, plants have to be handled. Cutting is generally done by hand, a light chopper or a sickle being used; where kale is fed in yards it is better to cut the stems into lengths of about 4–6 in., because otherwise many are pulled by the cattle out of the cribs, and are trodden underfoot. A good case can be made for a wide extension of the acreage devoted to kale, particularly now that milk production has assumed paramount importance in British farming. Kale can be grown on all soils, it responds well to fertilisers, singling is not essential, it is an admirable smother crop and, compared to the true root crops, it gives a high yield of protein per acre; but the expense of cutting it by hand and carting it daily to cattle is wellnigh prohibitive. Unsingled kale can be cut satisfactorily with a grass mower and the machine does not seem to suffer any serious harm; during the winter the cut plants do not wilt quickly and so a week's supply can be cut at once. This reduces the cost considerably but where cattle can be folded on the growing crop a much greater economy can be effected; many farmers now fold their dairy herds on kale, using electric fencing for the folds.

Swedes, turnips and kohlrabi are also often folded off on the field, but it is quite common for them to be clamped; though these crops (except white turnips) are regarded as winter hardy, a proportion will decay during a hard winter, whilst it was the practice to pulp them both for sheep and cattle, and where this practice persists handling the roots cannot be avoided. If these crops are intended for feeding to cattle they may be clamped in the manner described for mangolds, a few minor differences being involved. As the crops are more hardy to frost clamping may be delayed until later in the winter, if labour is wanted for other operations, though it should be completed by Christmas; kohlrabi grow slowly and mature late, and for full growth in most seasons they should be left in the ground until the turn of the year approaches. Smaller roots mean smaller air spaces (though an equal total air space) and hence more restricted ventilation in the clamp; consequently, where a long triangular clamp is built its base and height should be rather smaller than for mangolds, typical dimensions being 8 ft. by 4½ ft. It is very common to see crops with four rows of kale and four rows of turnips or swedes alternating; these are grown for folding by an arable sheep flock from December to March, and a frequent practice is to clamp the swedes in small pies scattered over the field. Each pie holds about 1 ton of the roots and is carefully built in conical form, with a covering of straw and earth; as the sheep fold off the kale the pies are opened, and the roots thrown whole into the field, or pulped and fed in troughs.

Occasionally swedes, turnips or kohlrabi are covered with a plough furrow to protect them from frost. The usual method is to turn one furrow over on to the top of each row of plants, taking care to avoid moving the latter, because it is believed that they keep better if undisturbed. On the other hand, some farmers pull the roots and lay four rows together, ploughing a furrow over the composite row from each side; in either case the roots are easily exposed when required, by harrowing the field.

POTATOES

Early potatoes are dug as soon as there is a satisfactory bulk of tubers from June or July onwards, according to district, soil, date of planting and variety. What the grower seeks is the greatest return per acre, and this of course is the product of the yield and price per ton; if the latter be high enough it is good business to 'murder' a crop by digging when the yield is only of the order of 2 tons per acre, and weight of tubers is increasing rapidly. If the market price is not satisfactory the farmer may decide to leave a crop of earlies to grow to maturity, and then to lift and clamp it. Potatoes will not keep in a clamp if they are lifted before they are mature, which state is shown by the dying down of the haulms, and which may be tested by rubbing the tuber; the skin of an immature tuber will crinkle up in front of the thumb, but when the tuber is ripe the skin is firm and will not break under the thumb pressure. Mid-season and main-crop potatoes are lifted during September and October, there being very considerable differences between varieties in regard to time of ripening. It is desirable that the operation should be completed by mid-October, because of the likelihood of wet weather being encountered later, and of the fact that tubers that have been frozen will not keep; where wireworm are present early lifting is very important as the longer the tubers are left in the ground the more will they be holed. The presence of blight may also affect lifting date. Spores fall from infected potato leaves and if they come in contact with a tuber that tuber will rot in the clamp. Good earthing up of the rows in July gives fair protection to the tubers, though heavy rain may wash the spores down through the soil. In a fairly dry year it may be satisfactory to put off lifting until the tops have died right down, though with some varieties this may not be until some weeks after the skins of the tubers have become firm; but in a wet period many spores will be washed into the ground and serious loss of tubers will result. It is generally considered best to remove the tops before an attack of late blight has developed. In some years this may have to be done as early as mid-August but generally it is a September job. For very small areas of potatoes the tops may be cut off and removed but for farm crops the method is to spray; sulphuric acid is very efficient for this purpose but it is expensive. Sodium arsenite is much cheaper and is used by some farmers; this compound is highly toxic to humans but as a spray for killing potato haulm it is effective and apparently safe.

DIGGING

The following are the methods of digging used in this country:

(*a*) *By hand.* A four-pronged fork is used and one man digs whilst another (more often a woman) picks up the tubers and fills baskets with them. Payment is usually by piece-work and based on the expectation that six or seven pairs of workers will clear an acre a day, other men being required to empty the basket into carts. This method has been very common, Irish labour being frequently employed, but the method is now too expensive for main-crop potatoes. Digging by hand, however, is still used for earlies, the price of which may be sufficiently high to justify more extravagance in handling; it may, indeed, be the only possible method with earlies whose large haulms may block a machine, and whose tubers are very liable to suffer from bruising by other methods.

(*b*) *By potato-digging plough.* This implement resembles a double-mould board plough (is, in fact, often converted from this) with the mould boards removed and their place taken by a series of bars slanting upwards and backwards; as the plough is taken along, the share undercuts the ridge which rides the bars, and the principle is that the soil falls through the bars but the tubers do not, and so are left lying on the top of the ground. But in practice many of the tubers are not uncovered and all the plough does is to open out the ridge, and pickers have to feel in the loose soil for the tubers; it is true that most of them will be only thinly covered, but if the soil is dry and contains many stones pickers go home with very sore fingers. The chief advantage of the potato plough over the spinner is that it does not damage the tubers so much, and for this reason it is commonly employed; another advantage is that if there is much green haulm the plough is less likely to be blocked by it than is the spinner.

(*c*) *By the spinner.* This implement carries a broad share that runs at an adjustable depth below the ridge, which it undercuts. Revolving arms throw the ridge sideways and spread it out, the tubers tending to travel farther than the soil, and being left on the surface of the ground. Undoubtedly the spinner exposes the tubers much more efficiently than the potato-digging plough, but it has the drawbacks mentioned above. The revolving tines hit the tubers at considerable speed, and in some cases the bruising has a serious effect on the keeping quality; watery wound rot may develop and may render a large proportion of the crop unsaleable. It is possible to obtain rubber sleeves to put on the tines, and these have a useful effect in reducing bruising. It frequently happens that the spinner cannot be used because there is too much haulm present, which becomes entangled in the revolving arms and blocks the machine; in some cases, when the ridge is very massive, the spinner has to be discarded as the arms cannot move the weight of soil in the ridge. In dry, friable soil, however, the spinner does very good work, and if the tubers are fully mature they are undamaged by its action. Both the spinner and the potato-digging plough are seen commonly in potato-growing districts; it is probably true to say that the spinner is less popular than it was, because of the fear that it may reduce the keeping quality of the tubers.

(*d*) *By elevator digger*. In this type of implement a broad share undercuts the ridge which rides up a chain of parallel bars; the soil falls through the bars and the potatoes are delivered in a neat row on top of the ground. In good conditions the elevator digger is definitely preferable to the spinner and pickers who have worked behind the former sometimes refuse to follow the latter. As the autumn advances, however, and the ground becomes stickier, the elevator digger becomes mudded up and growers with many acres expect to have to change to the spinner about the second week in October. There is a strong tendency at the present time towards the elevator type.

As with sugar beet, so with potatoes, new and larger machines designed to eliminate hand labour are being produced; in the case of potatoes, harvesting machines have been available for some time, but only in recent years has perfection of design been approached, and as yet they play little part in this country. Although it is to be anticipated that in future these machines will usually find a place on the larger potato-growing farms, there are certain obvious factors which militate against them. They will, necessarily, be expensive to buy and heavy in draught, and it will be difficult to produce one which will not damage tubers; one grave disadvantage lies in the inability of a machine to distinguish between a tuber and a clod or stone of similar size and shape. Already there are machines which do a satisfactory job in ideal working conditions but such conditions are rarely encountered. Much ingenuity has been shown in designing potato harvesters but the progress made has not been so great as with sugar-beet harvesters.

The harvesting of the potato crop necessitates a large labour force, and hence it is important to organise the work so that wasted time may be reduced to a minimum. Where the spinner is used the outside rows of the field are best removed a day before starting the main work, because the spinner may throw the tubers into a ditch; a plough is used for these rows. The method is to take a strip of thirty or forty ridges at a time and work up one side of the strip and down the other. All the tubers exposed at one passage of the spinner must be picked up before the next passage, or they will be covered when the next ridge is thrown out; with a well-balanced gang the spinner or plough keeps going all the time and the pickers are fully employed, but the varying industry of the latter usually means that some have to wait while the man on the spinner helps the laggards. With a normal crop twelve to sixteen pickers are required to keep pace with the spinner, half of them working on each side of the strip being cleared; it is best to divide the length of the ridge into equal distances and to mark these with pegs, so that each picker knows exactly how far his responsibility extends. Women are employed very frequently as pickers, their daily wage being less than that of men, despite the fact that, for anatomical reasons, women do the work better and more quickly. During the War of 1939–45 schoolboys gave yeoman service in this work and it is certain that many acres of the war-time potato crop would never have been cleared but for their assistance; frequently, however, the gangs of schoolboys were too large so that there were idle periods, the results of which can readily be imagined, with so many suitable missiles ready to hand. With the passing of the war emergency local education authorities refused to allow children to stay away from school for this purpose so that this valuable source of labour has dried up. Pickers are supplied with baskets of about 1 bushel capacity, and as these are filled they are left for the carters to empty. Two or three horse carts are necessary to keep up with the work of emptying baskets, the number depending on the rate of working and the distance of the clamp; there must be a man with each cart, the work of lifting and emptying the baskets into a cart being heavy. Horses are now rarely available and usually there is only one tractor for the work; with two or three men to empty the baskets into the trailer this one tractor may be able to keep up with the picking gang if the clamp is handy to the field. When a part of the field has been covered it is harrowed with a two-horse or a drag harrow. This is done for two reasons—first, to collect the haulms into heaps for burning (the haulm normally carrying much disease), and secondly, to expose tubers left by the spinner or plough. After the harrowing a second picking is carried out, the pickers working across the field in a line, with a cart following them so that baskets may be emptied as they become full. Often a second harrowing and a third picking are carried out, the work being repeated as long as there are sufficient tubers exposed to justify it; in some cases it may even be worth while to have a final picking after the field has been ploughed in preparation for the next crop. It is not only the value of the potatoes collected that makes the pickings desirable, as it is important to avoid, as far as possible, leaving tubers on the field; these ground keepers, as they are called, give plants which are weeds in the subsequent crop, and they serve as hosts to pests and pathogenic fungi attacking potatoes, and so reduce the benefit of resting the field from that crop.

CLAMPING

Potato clamps are nearly always triangular in cross-section but are much narrower at the base and lower in height than those of mangolds; the relatively small size of the potato tuber, and the great number of diseases to which the crop is prone, make it important that the base shall not exceed 7 ft. in width, and the vertical height not exceed 4 ft. The site must be chosen so that there is no fear of water running into the bottom of the clamp, and in close proximity to a hard road; it is better not to dig anything resembling a trench but just to level the ground and put the potatoes on the surface. Sometimes a thin layer of straw is put beneath the clamp, with the object of keeping the tubers dry, and in the hope of improving the ventilation. The clamp is carefully built and covered with straw as in the case of mangolds; earthing up should not be carried out until a fortnight after the clamp is completed, and then is usually done so that the whole is covered except for wisps of straw every two or three yards along the apex. The man building the clamp must keep a sharp look-out for diseased tubers, which should be kept apart from the healthy ones.

In seed-growing districts it is common to make a number of very small clamps and to scatter them all over the field as it is cleared. This has advantages where few workers are available as no carters are required, the pickers emptying their baskets straight to these clamps; in those areas wheat rarely follows potatoes and as the tubers are sold as seed as fast as they can be riddled the little clamps are cleared in good time for sowing a spring cereal. A variety of novel clamping methods has been employed in recent years. Pig-fattening houses of the Danish type have been used very successfully, the separate pens being filled up to 4 ft. deep with potatoes and covered with a good depth of loose straw; one great benefit gained is in the riddling which can be done under cover. Rectangular clamps have been built between walls of straw bales. The width should be no more than 10ft. and posts should be driven into the floor of the clamp every 3 or 4 ft. of its length, and each post encircled with a column of drain-pipes to provide a ventilation chimney. A height of fully 4 ft. may be allowed and the clamp should have a generous covering of loose straw, the whole being finished with a roof of thatch. If the potatoes are free of disease, dry and clean when lifted they keep well in such clamps. The tendency is to change from the traditional field clamps to storage under cover and this will probably be the general method of the future. Large growers are equipping themselves with special storage buildings in which the potatoes are piled 8 ft. or more high.

MARKETING

If the price is good many potatoes are sold at the time of digging, and it should always be the aim to dispose of the crop as quickly as possible if any appreciable amount of blight was observed during growth; when a blighted crop is left for long in a clamp the loss through rotting may be very serious indeed. In the past potato prices exhibited violent fluctuations from year to year and even during one winter, so that the financial success of the grower depended to a large degree on his astuteness in marketing. With the advent of the Potato Marketing Board in 1934 price fluctuations were considerably smoothed out and from 1940 onwards the Ministry of Food undertook to buy all the potatoes produced, at prices fixed well ahead and varying with variety, soil, district and month of sale. Now there is again a Potato Marketing Board with price support from the government which guarantees a minimum price for sound potatoes. The price rises throughout the marketing period which extends from September to June, when new potatoes, coming from other countries or from early districts, render old ones unsaleable. To a farmer with a large weight of the previous year's crop still unsold a severe frost in late May is a godsend; it will postpone the date when earlies will come on the market, and consequently it may cause the price of old potatoes to soar. In such circumstances it may be possible to dispose of the remainder of the old potatoes (the sale of which was becoming problematical) at prices which would have seemed fantastic before the late frost, and buyers will not be particular if a few of them are bad. It is no use riddling potatoes when they are lifted because too often many will go bad in the clamp; they are riddled at the time of sale, bad ones being picked out during the process. The tubers which are saleable as a vegetable are known as 'ware' potatoes, and normally these will not pass through a riddle containing holes $1\frac{5}{8}$ in. square; the size, however, is variable, the Potato Marketing Board having the power to raise or lower it if they foresee a glut or shortage. The intermediate-sized tubers are known as 'seed' potatoes, this size being used as setts; it must be realised, however, that only 'seed' size potatoes from a crop grown from Scotch seed should be planted, others often being fed to stock, or in times of shortage being saleable as a vegetable. 'Seed' potatoes pass through a $1\frac{5}{8}$ in. (or near figure fixed by the Marketing Board) riddle but remain on one with holes $1\frac{1}{2}$ in. square (occasionally the latter is reduced to $1\frac{1}{4}$ in.). The small potatoes which pass through a riddle on which 'seed' potatoes remain are called 'chats', and should be fed to stock though often they are allowed to rot down in a heap.

The simplest form of riddle is a sieve standing on a stick; the potatoes are put on the sieve which is shaken. A better type, and one commonly seen in use, consists of a framework carrying two sieves one above the other; tubers are shovelled on to the top one of the sieves, which are then shaken. The tubers remaining on the upper sieve are carefully inspected for diseased ones, which are picked out by hand, and then the remainder are tilted into a sack; 'seed' tubers are delivered to one side, whilst chats fall through the lower sieve to the ground. Growers of large acreages have continuous working machines driven by a petrol engine. One man with a potato fork (which has long curved tines tipped with knobs) puts the tubers into a low hopper from which an elevator takes them up to the riddles. The ware potatoes which stand on the top riddle pass to a conveyor which delivers them to sacks, the conveyor being 3 or 4 ft. long; two or more workers pick diseased or damaged potatoes off as they travel along the conveyor. A gang of 6 or 7 workers dealing with a good lot of potatoes can get out 12 or more tons of ware potatoes in a day, sacking them up in 1 cwt. sacks, the mouths of which are roughly sown up with strong twine.

The proportions of the three sizes which are obtained are extremely variable, and figures are only given with the greatest hesitancy. Favourable growing season, absence of disease and rich soil are three important factors tending to increase the proportion of ware; there are also marked differences between varieties in this respect. The following would be regarded as normal figures:

Ware	75 per cent
Seed	15 per cent
Chats	10 per cent

It must be understood that these proportions refer to the weight of crop at the time of riddling; during the time potatoes are in the clamp there is some loss (probably of the order of 5 per cent) through respiration and there may be a ruinous loss from diseases, of which blight is the chief offender.

CHAPTER XII

COSTS

It is possible to develop a very pretty argument as to whether farming is an art or a science; cogent points may be made in support of either view, but in reality it is neither the one nor the other—it is a business. The supreme consideration is the economic one so that it would appear a glaring omission if, in a book on crops, no mention were made of the costs of production of these crops. But this matter bristles with difficulties. Thirty years ago many figures were published which purported to show with great precision how much it cost to produce a quarter of wheat or a ton of potatoes, but there was always much disputation among agricultural economists over the manner of calculation and now it has become evident that such costs have very limited value.

When full cost accounting was in favour the aim was to arrive at average costs over a group of farms, deriving the data from time sheets, receipts for materials and so on; but an average is at best a lifeless abstraction of limited utility and when applied to individual farms it becomes a mere absurdity. It is, of course, possible to perform the necessary arithmetic and arrive at average figures from farms with soils varying in type and fertility, of different sizes and situations and with differing organisations; but when the work is done the result can clearly apply to no one holding in the country. This elementary consideration limits the objectives of cost accounting to the provision of figures typical of certain specified conditions.

The expense of any operation is largely a matter of man and horse or tractor labour, and at first sight it might appear that the cost of the first of these, at least, was capable of accurate assessment. It is true, of course, that a figure can be given for the cost of employing a man for a day, but it by no means follows that this is a fair price for the task he accomplishes in a day's work. The point was stressed in Chapter I that no crop can be regarded as a separate entity; it may be good business to introduce a crop into a farming system because it creates a demand for labour at seasons when it would otherwise be difficult to find productive work for the men. The crop may incur heavy apparent labour charges, but clearly it cannot be debited fairly with the whole of these, since it helps the other enterprises of the farm by enabling the farmer to keep a more constant staff throughout the year. Similarly, the true cost of an operation must vary with the season of the year in which it is performed; ploughing during the slack winter months has a lower real cost than ploughing in a busy September, even though the labour employed and rate of working may be the same. It is only possible to record considerations of this nature, because there is no conceivable basis on which differential costs for the same work can be approached.

Even if the simplest case—that of a farm producing only one commodity—be postulated, a large element of doubt remains. To calculate all the outgoings of such a farmer, and to derive from them his cost of production, ignores any effect, beneficial or harmful, which his system may have on the fertility and freedom from disease of his land. Gain or loss of fertility as a result of the year's working may be an important item of very considerable value, but it is one to which no figure can be put. There is very great uncertainty over the residual values of manures, and farmyard manure, in particular, involves heavy expense which is quite capable of influencing very materially the final calculated cost of a crop; there are valuation tables suggesting how the cost may be spread over successive crops, but the figures in the tables are only guesses, the intelligence of which is unproven and, indeed, open to some doubt. Farmyard manure is one of a number of by-products which may arise on a farm and which a good organisation will utilise to the utmost. An enterprise may be introduced into the system chiefly as a means of disposing of a by-product, which otherwise has little or no value; what should be charged to the enterprise for the by-product is a difficult matter to settle.

It is quite easy to imagine the endless arguments among agricultural economists when they were engaged in cost accounting, but it is not proposed to delve more deeply into this question, which cannot be dealt with summarily and which, in fact, lies rather outside the scope of this book. However important monetary considerations are, farmers will not be helped by figures which are untrustworthy, as would be the costs per acre calculated on questionable premises. All that is proposed in this chapter, therefore, is brief discussion of the cost of employing labour, horses and tractors, of rates of working which may be expected and a slight reference to the help which a farmer may get from an analysis of his financial results.

THE COST OF MAN, HORSE AND TRACTOR LABOUR

Since the First World War there have been Wages Boards to fix minimum wages for agricultural work and their findings have the force of law; these Boards have had many vicissitudes, but the present position is that there is one Board for the whole of England and Wales with the duty of fixing, from time to time, minimum wages for men and for women, and for boys and girls according to their age. Practically every conceivable detail is covered—the hours to be worked, overtime rates (with special rates for Saturday afternoons and Sundays), holidays to be allowed and maximum deductions for rent of cottage. A farmer is forced to pay at least the minimum (even if his men are on piece-work their total earnings for a full week must be up to the minimum) as otherwise he is liable to a heavy fine in addition to all back deficiencies. Some farmers are so conscious of the importance of having good men that they pay a little more than the minimum, but many cannot afford to do so; piece-work payment, and bonuses are their only chances of providing incentives for good work.

At the time of writing the minimum wage for men aged 20 and over is £7. 1*s*. for a 47-hour week. The cost per hour therefore appears to be 3*s*. or, since the week consists of 5½ days, the cost per day 25*s*. 6*d*. The real cost, however, is considerably higher. Overtime rates are 4*s*. 6*d*. an hour and general farm workers average 3½ hours a week overtime; specialised workers, including tractor drivers, do rather more overtime than this. Workers are entitled to 12 days holiday a year and to six Bank

Holidays, all with pay. To actual wages must be added the farmer's weekly contribution to the National Health and Unemployment Insurance (6*s.* per week) and 5*d.* per week as premium to cover Employer's Liability. Then many workers have cottages rent free or for a nominal rent, and various perquisites such as milk, eggs, land for growing potatoes and so on. It is clear that the real charge to the farmer for a man-day is considerably above the minimum weekly wages divided by 5½, and it might indeed be taken as 35*s.* Before 1940 the Wages Boards also fixed special rates of pay for the corn harvest, and all full-time workers expected to be 'given a harvest', which meant a welcome addition to their annual income; nowadays wages for harvest are the same as for other work, but a man willing to work overtime can count on the opportunity to get a useful increment to his wages during harvest.

Numerous estimates have been made of the cost of horse labour. This is complicated by the fact that the total cost of keeping a horse for a year rises with increase in the amount of work done, since he must be fed more, whilst the daily cost falls, since this is obtained by dividing the annual cost by the number of days worked. It was common to include under horse expenses, charges for the purchase and maintenance of ordinary tillage implements, waggons and carts; this item was generally taken to be about one-quarter of the whole cost of horse labour. But the status of the horse on all except the smallest farms has undergone a radical change in the last 25 years; tractors now do the heavy work and many a horse leads a life of ease, which makes what little work he does extremely expensive. Few deny the value of the horse for odd jobs, or for anything which involves much standing about, and farms which are highly mechanised often have one or two horses for such work; taking the farm as a whole these horses are well worth their place, but any computation must give a very high figure for the cost of a horse-day. The yearly cost of keeping a horse is in the neighbourhood of £60, which figure includes nothing for the implements he uses since, in fact, he rarely uses them. Food accounts for some two-thirds of the £60, the remainder covering veterinary expenses, attendance and depreciation; the last of these items is extremely variable, as some farmers breed their own horses and sell them at their prime, so that there is really an appreciation in value during their working life on the farm. If a horse costs £60 a year and works 200 full days the charge for a horse-day is clearly 6*s*., but from farm to farm the range of cost must extend at least from 3*s*. to 12*s*.

There are also many uncertainties in calculating the cost of tractors. In this case there is bound to be an important item for interest and depreciation, the size of the item varying inversely with the number of days in the year on which the tractor is worked. A tractor is generally considered to be fully employed if it does 100 days' work in a year; for that the annual cost is approximately £150, giving a figure of 30*s*. for a tractor-day. Of the expenditure on tractors, approximately half is for fuel, one-third is for interest and depreciation, the remaining one-sixth covering repairs and service. One of the great advantages of a tractor is its ability to work long hours when the occasion demands, so that it is perhaps better to think in terms of cost per hour rather than per day. A 35–40 h.p. wheeled tractor, working 800–1000 hours a year, costs about 4*s*. to 4*s*. 6*d*. an hour, exclusive of its driver's wages. With heavy tractors the cost rises more than proportionately to the power and this is particularly true of tracklayers, whose tracks are very expensive in renewal; thus a 35–40 h.p. tracklayer would probably cost 6*s*. an hour and one of over 50 h.p. around 10*s*. an hour.

RATES OF WORKING

No true picture of crop production can be obtained without knowledge of the labour and power required for the various operations, and Table IX is designed to give some guidance on this question, which is of obvious importance in farm management. It must be emphasised that the figures are no more than guides, for in practice there is very great variation. The amount of work that horses can do in a day is limited by their endurance; they are slow to get out to the field and some time is needed for attendance on them in the stable, so that 6 full hours' working is all that can usually be managed in a day and, on heavy work, all that the horses can sustain. With tractors, servicing is a time

TABLE IX. RATES OF WORKING

A. *Horse or Tractor Work*

	Acreage covered per day with	
	Horses	Tractor
Ploughing (per furrow)	1	$1\frac{1}{2}$
Cultivator, heavy harrow or disk	7	12
Light harrow	12	40
Roll	10	30
Ridging or splitting ridges	3	12
Drilling	10	20
Spreading chemical fertilisers	10	15
Potato planting	—	7
Single-row hoeing	$2\frac{1}{2}$	—
Multiple-row hoeing	8	12
Cutting corn with binder	10	20
Combining (per foot of cut)	—	$1\frac{1}{2}$
Raking	15	25
Lifting potatoes	2	2
Harvesting potatoes	—	3
Lifting sugar beet	2	7
Harvesting sugar beet	—	2

B. *Gang Labour*

	Men	Horses	Tractors	Acreage covered per day
Carting farmyard manure (12 tons per acre)	5	3	—	3
Carting farmyard manure (12 tons per acre)	3	—	1	1
Potato planting and covering	8	1	1	6
Potato lifting and clamping	20	2	1	$1\frac{1}{2}$
Potato riddling	5	—	—	$1\frac{1}{2}$
Carting mangolds from field heaps	5	3	—	2
Carting and stacking corn	7	3	—	10
Carting and stacking corn	3	—	1	4
Carting and stacking corn	10	—	3	15
Threshing corn	11	1	1	12
Sweeping and stacking straw	8	—	1	17
Pick-up baling and stacking bales	7	—	2	12
Filling silo: young grass in pit	4	—	2	6
Filling silo: oat and tare in tower	7	—	2	4

C. *Manual Labour*

	Acreage covered per man per day
Spreading farmyard manure	$1\frac{1}{2}$
Hoeing: corn	$\frac{1}{3}$
Hoeing: potatoes	$\frac{1}{2}$
Hoeing: sugar beet (singling)	$\frac{1}{5}$
Hoeing: sugar beet (seconding)	$\frac{1}{2}$
Stooking corn	4
Thatching	5
Pulling, knocking and topping sugar beet	$\frac{1}{6}$
Pulling and lumping mangolds	$\frac{1}{3}$

consuming necessity, fuel supply has to be organised and breakdowns are not unknown; long hours are often worked but the figures in the table only refer to normal days, where no overtime is earned by the driver. The condition of the land has a big effect on the acreage covered by an implement. Rate of working is reduced on heavy land, particularly when it is wet. Horses with their feet encased in mud can do no more than plod slowly along, whilst tractors have to contend with a heavier draught, suffer from wheelspin and occasionally get bogged. Size and shape of field, as influencing time wasted in turning, have important effects on acreage worked. For horses, the various implements are fairly standardised in width, but for tractors there is considerable variation. It is important where draught is small, as with light harrows, that the working width should be sufficient to keep the tractor well loaded, lest it runs too cool with consequent dilution of the sump oil; in any case it is clearly uneconomic to use a narrow width, because then a large proportion of the tractor's power would be used to propel itself and the driver's wages would have to be spread over few acres.

In §A of Table IX, potato planting refers to planters setting three rows at once and potato lifting to single-row spinners or elevator diggers; with potato lifting some allowance must be made for delays caused by slow pickers, which factor need not be considered with a complete harvester. In regard to sugar beet, horse-lifting ploughs only take one row at a time, whilst tractor tool-bars carry three or four lifters; the complete harvester will cover fewer acres per day than will the potato harvester, since sugar-beet rows are nearer together than potato ridges. With both these harvesters rate of working may be expected to rise as models of greater mechanical perfection are developed. Tractors vary in power and speed, but for horses one general guide can be useful. It is a good day's work to plough an acre a day with a single-furrow plough, turning a furrow 10 in. wide; for any other horse implement an approximate estimate of a day's work can be obtained by dividing the width of the implement in inches by ten.

The organisation of gang labour, the number of workers, horses and tractors available, show great diversity from farm to farm; certain organisations have been selected for §B of Table IX, but the reader should be quite clear that these are only illustrations and that many other possibilities exist. During the war years large gangs were provided by County War Agricultural Executive Committees but it by no means follows that a gang twice as large does a job in half the time; the ideal to aim for is a well-balanced gang, just large enough to make the work continuous, and with women or children for the lighter tasks. In regard to potato planting, the illustration is where a tractor with tool-bar and three ridging bodies opens and splits the ridges, one man takes out the tubers with a horse and cart and generally supervises, whilst six planters (very often these are women) drop the potatoes on an acre a day each; if farmyard manure or chemical fertiliser is applied in the ridge, additional labour is required. The potato-lifting gang cited is for working with a spinner or elevator digger and should clear the 1½ acres a day, including the second picking after harrowing; of the workers, one will be driving the tractor and spinner, one building the clamp and strawing it, two with horses and carts or with a tractor and trailer collecting from the filled baskets and taking the tubers to the clamp, the remaining sixteen being pickers. Strenuous efforts are now being made to mechanise the making of silage. In the first illustration it is assumed that a 'cutlift' is used, and in the second that the crop has been cut and tied with a binder.

Even with manual labour, very wide variation is to be found in acreages covered. Speed of hoeing, for instance, must depend on the number and type of weeds present, the nature and condition of the ground; good row-crop work, up to within an inch of the actual row, and a thin seeding rate, speed the work of singling. Standard of labour is a variable of obvious importance; piece-work rates for similar work vary from farm to farm by as much as 50 per cent.

FARM ORGANISATION AND BUDGETING

Technical efficiency in crop production does not necessarily ensure profitability in farming; the days when a farmer had merely to 'do his duty by the land' to achieve success are past and now he cannot rely on his unaided judgement but must think in terms of figures. But, for reasons advanced earlier in this chapter, true costs of production for the different crops are impossible to obtain; because of the interrelation of the different enterprises in a farming venture, the economics of crop production is not a part of crop husbandry but of farm management.

Advisory Economists now render assistance to farmers by providing them with standards calculated from similar farms, standards by which their own efforts can be judged. Distribution of the acreage of the farm over the various crops, yields of crops and from livestock, receipts per 100 acres or per £100 labour costs—figures such as these can be determined without making any assumptions, and a farmer can find the weaknesses of his own organisation by comparing his figures with those from other farms. It is not proposed to pursue this important point further but no apology is offered for returning here, at the close of the book, to a statement made on p. 1, namely, that no crop should be regarded as a separate entity; in a country of mixed farming this is fundamentally important. A farmer may, for instance, be well satisfied with the cheques he receives month by month in the autumn for sugar beet delivered, and conclude that they amply cover any expense he has incurred in growing and harvesting the crop; there are, however, other considerations to be taken into account before he should decide whether the crop has a rightful place in his farming system. On the one hand he may be underestimating the value of the crop to him, because its by-products (tops and pulp) may have enabled him to keep more profit-making cows; on the other hand, the heavy labour demand for the beet at lifting time may have put his other autumn work seriously behind schedule, with consequent losses on other crops. Before making up his mind to introduce or abandon a crop, to buy a new implement or to change his system in any other way, the farmer should try to estimate the effect of the alteration on the financial results of his organisation as a whole; there may be all manner of repercussions, and before committing himself he should assess the collective effect of these on that for which he is farming—his profit.

INDEX

INDEX

INDEX

Printed in the United States
By Bookmasters